国外高等院校土建学科基础教材（中英文对照）

BASICS

设计与居住

DESIGN AND LIVING

［德］扬·凯博斯　编著

马　琴　万志斌　译

U0725529

中国建筑工业出版社

著作权合同登记图字：01-2007-3332 号

图书在版编目（CIP）数据

设计与居住/（德）凯博斯编著；马琴，万志斌译. —北京：中国建筑工业出版社，2011

高等院校土建学科双语教材（中英文对照）◆ 建筑学专业 ◆

ISBN 978-7-112-11584-6

Ⅰ. 设… Ⅱ. ①凯…②马…③万… Ⅲ. 建筑设计-高等学校-教材-汉、英 Ⅳ. TU2

中国版本图书馆 CIP 数据核字（2009）第 209821 号

Basics：Design and Living/Jan Krebs
Copyright © 2007 Birkhäuser Verlag AG（Verlag für Architektur），P.O. Box 133，4010 Basel，Switzerland
Chinese Translation Copyright © 2011 China Architecture & Building Press
All rights reserved.
本书经 Birkhäuser Verlag AG 出版社授权我社翻译出版

责任编辑：孙书妍
责任设计：郑秋菊
责任校对：刘 钰 关 健

高等院校土建学科双语教材（中英文对照）
◆ 建筑学专业 ◆
设计与居住
［德］扬·凯博斯 编著
马 琴 万志斌 译
＊
中国建筑工业出版社出版、发行(北京海淀三里河路 9 号)
各地新华书店、建筑书店经销
北京嘉泰利德公司制版
北京建筑工业印刷厂印刷
＊
开本：880×1230 毫米 1/32 印张：3¾ 字数：120 千字
2011 年 5 月第一版 2018 年 9 月第二次印刷
定价：**24.00** 元
ISBN 978-7-112-11584-6
　　　（32399）
版权所有 翻印必究
如有印装质量问题，可寄本社退换
（邮政编码 100037）

中文部分目录

CONTENTS

序

　　居住往往是建筑学需要解决的头等大事。即使准建筑师们一开始会觉得凭借他们对自己家的经验来设计住宅是一件很容易的事情，但是他们必须把自己的想法放在一边，然后通过对新概念的分析解决住宅的问题。在设计起居空间的时候，最大的困难在于我们不能从自己和自己的需要出发，而必须在头脑中牢记普通住户，因为他们的经验和价值观会与设计师的想法大相径庭。因此，居住建筑必须体现出使用者能够对自己的空间有所改进并且根据他们自己的想法创造生活品质的特点。这个特点不仅适用于一般的住宅环境；对于单个起居空间或者居住者认为重要和舒服的空间的设计来说也是如此。因此对于建筑师来说，努力创造家庭的空间品质是一件很重要的事情。

　　这套"国外高等院校土建学科基础教材"系列逐步建立了一个全新的行为领域中的重要原则，为建筑学的研究提供了一个有效而实用的工具。它们不是专业知识的全集，而是面向学生的深入浅出的阐述，以促进他们对不同学科领域中重要问题和要素的理解。

　　本书没有列举那些随处可见的实际建成的例子。也是出于同样的原因，本书也没有提供理想化或者极端化的起居空间设计方针，因为居住空间总是居住者个体和主要社会或者气候状况的反映。相反，本书对基本概念进行了解释，阐述了如何把单个的空间连接起来，以及建筑形式的变化方法。本书的目的是让人理解正确创造卧室、起居室和整个居住单元的方法，并且能够用这些知识全面而慎重地进行自己的住宅设计。

　　本书想要帮助读者找到一个把复杂的住宅设计简单化的办法，强调居住的基本内容。"基本内容"一章描述了概况，并举了一些设计方法的例子。"居住元素"一章讲的是家的不同用途和功能，以及基本的和特殊的要求。最后，本书从城市的角度对基本方法以及它们对家庭生活的影响进行了分析，并且在"建筑形式"一章中进行了讲解。

编者：贝尔特·比勒费尔德

Dwellings are usually the first design projects to be tackled in architecture courses. Even if would-be architects find their way around this area easily at first because of their own experiences of homes, they have to put their own ideas aside and address the topic of housing through the lens of new concepts. When designing living space, the challenge lies in not starting with oneself and one's own needs, but in having the eventual occupants in mind, as their horizons of experience or values can be very different from the planner's. So dwellings must be endowed with qualities that permit the users to develop in their own space and create lives for themselves according to their own ideas. This does not apply to the general home environment alone: the design of individual living areas and spaces is crucial if the occupants are to feel at home and comfortable. So it is important that architects should make deliberate and intensive efforts to create spatial quality in homes.

The "Basics" series of books aims to present the fundamental points about a new field stage by stage. Instead of compiling an extensive compendium of specialist knowledge, it aims to provide students with explanations that are readily understood and give insight into important questions and parameters of the various fields.

This "Design and Living" volume therefore refrains from providing built examples, which are already exhaustively available. For the same reason, it does not offer idealized or polarized guidelines about how to design living space, as dwellings always gives an individual picture of their user and the prevailing social or even climatic conditions. Instead, "Design and Living" explains fundamental concepts, the way the individual living areas can cohere, and building form variations. The aim is to understand approaches to making rooms, living areas and entire residential units fit together, and to be able to use this knowledge to develop one's own housing construction design work competently and prudently.

This book is intended to help readers to find an easy way into the complex subject of housing design, emphasizing fundamental aspects of living. The chapter Basics presents general conditions and provides examples of design approaches. The chapter Elements of living describes the various uses and functions of a home, along with general and special requirements. Finally, the book examines the fundamentals of approaches to the urban context and their impact on home life, and the fundamentals of access in the chapter Building forms.

Bert Bielefeld, Editor

INTRODUCTION: WHAT IS A HOME?

The term "home" covers a large number of needs and requirements. A home requires different areas, including those for sleeping, cooking, eating and hygiene, in a single location. Home life is something quite ordinary in this context, but by no means unimportant. Life at home is one way people have of defining themselves. The place a person chooses to live identifies their preferences and constraints. The size, form and furnishings of a home, and the way it is occupied, influence the states of mind of the people in it. It is a place for retreat, but also a place for communication; for both intro-verted and extroverted situations. Fundamentally different sets of events may take place at the same time, and it is not always possible to combine the various interests and functions without making compromises.

Homes are also linked with constantly advancing personal change. This is particularly clear in the way a person develops in our society: child-hood – youth – a period of education – a period of working and establishing a family – retirement – old age. These stages in life usually bring different needs with them, and thus changes in the living environment. The home is adapted to the changing life situation as much as possible. Such possibil-ities are limited. It is often simpler to move than to modify, extend or im-prove the current home while still living in it. There are homes for all walks of life and in all price classes, but searching for a home is seldom simple. We have fixed ideas about what we want for a home, and make concessions only for financial or time reasons.

Planning a residential unit should not only address the necessary functions and resultant building costs, but also fundamentally aspire to creating a higher quality of living. The occupant's needs are an important basis for the design. If plans are made on an individual basis, if the future occupant is known, his or her ideas can be considered directly and the home is tailor-made. Without a particular user in mind, for example when building rental homes or developing property, the planner has to work on the basis of the general requirements of a fictitious target group. Here it is usual to design the most flexible standard ground plans for different us-ers, leaving sufficient scope for individuality, but also accounting for the housing market and demand in the long term.

There have always been and still are different forms of housing all over the world. Regional developments depend on different climatic condi-tions, local features such as topography and material resources, and not least the general cultural situation. Thus, parameters in desert regions are different from those in temperate zones, cities have different demands from thinly populated areas, and so the world's many housing cultures are distinct from each other. This certainly means that many dwelling design criteria are site-specific, but some basic ones can be adapted for other places or at least broaden the designer's horizon and release the potential for ideas. The human being is the measure of past, present and future developments. In this context, looking back at the history of housing can offer numerous ideas and tried-and-tested concepts for current planning tasks.

The current situation in society and housing construction is a snapshot, and future developments are difficult to predict. Nevertheless, the use cycle for a residential building can extend over a number of decades. So we are not building for today's needs alone, but for future generations as well. Aesthetics are less decisive than the substantial quality of the buildings, their functionality and the possibility of varied uses in future. The needs and requirements of a particular society form the basis for housing appropriate to its times. This basis should be constantly examined to create a quality of housing that can persist for as long as possible.

There can be no full answer to the question posed at the outset, "What is a home?" as each person has his or her constantly changing ideas about the subject of home and housing, in addition to general requirements. For this reason, interested readers are invited to ask themselves what a home means, in order to develop their own concepts and find innovative approaches. The following chapters are intended to contribute to this.

Housing design creates sequences of events in life that are influenced by a large number of factors inside and outside the home. Important standards for comparison in this context are a person's basic requirements, the dwelling's particular surroundings, and not least the principles of the design approach. These essential considerations for every residential unit are summarized below.

LIFE CYCLES

Human life cycles

Many of the demands made on a home are influenced by the life cycle of human beings, who have changing needs and requirements as they age. If a dwelling's structure cannot be adapted to a new life situation, it could stop being viable, and the occupant will have to look for an appropriate alternative. It therefore makes sense to consider the various phases of life at the preliminary planning stage. An ideal home would be suitable for children, elderly people and disabled people, but of course not all these requirements can be met at the same time, nor do they need to be. For example, cost prevents having a lift in every building with more than one floor. But a number of realistic measures can be implemented sensibly, or at least prepared: for example, service equipment such as switches, power points, and door and window handles can be installed at a level that wheelchair users and children can also reach. Provision can also be made in advance for the necessary space that would allow modification of the home for elderly or disabled people. Features such as unnecessary height differences on a particular floor or very narrow doorways should be checked at the design stage, as aesthetic approaches can clash with later adaptations to a different life situation.

Barrier-free housing

Freedom from barriers is a particular condition for humane dwellings. Barrier-free construction is intended to make it possible for everyone to use a place or a building, regardless of their constitution. Restricted physical capacity should not mean that people are excluded. In this context, special barrier-free homes must be usable by everyone, and put their occupants in a position to live largely independently. People who are blind or who have mobility impairments, people with other disabilities, elderly people, and children benefit in particular. Not every home can and must be barrier-free in this sense, but options for possible modification can be left open.

> ◫

The building as a whole should also be seen in the context of life or use cycles: most residential buildings are used over periods of several decades, and are refurbished or converted at intervals so that there is no substantial loss of quality, and adaptations are made to changing needs. If these use cycles and the resultant periods of time can be made to match the occupants' life cycles, there are opportunities at the planning stage for measures that can be implemented sensibly during later use. In this way, the changing needs of the occupants can be planned for, not just with respect to long-term economic factors, but also to meet changing needs and necessary modifications.

ORIENTATION

One key quality criterion for a home is natural light. Large or small apertures for admitting light to a home make a considerable difference, and whether the light comes from north, south, east or west is another important factor. The energy plan for a building is also affected by the relative position of the sun. Solar energy is deliberately directed into rooms through windows according to the season and the climate zone, or blocked by sunshading devices to avoid overheating and dazzle. Tree and plant growth, sunshading installations, balconies and protruding roofs can be used for the latter purpose. But small windows or no windows at all may be sensible solutions under certain circumstances; the same applies to the opposite climatic conditions in cold regions, where a great deal of heat energy is lost through the windows on the north side of a building.

Local features such as buildings, roads and open spaces, as well as special topographical features and trees, influence the design for a dwelling

---

\\ Hint:

International and national standards contain instructions for designing barrier-free homes: ISO/T 9257 Building construction: handicapped people's needs in buildings – design guidelines

There are very few international and national standards relating directly to the creative design of residential buildings. For example, the Swiss standards (SN) contain no regulations of this kind, and other nations produce few standards, with the exception of provisions on function and safety. However, standards are subject to constant change, and international standards, in the guise of ISO (International Organization for Standardization) and EN (Euronorm) are working towards uniformity. Such standards, if they exist at all, represent only part of the necessary planning principles that can influence a housing design. Instruments such as local building regulations are similarly important, along with specific qualities of the materials intended to be used for realization, and not least the future user's requirements.

and determine the orientation of areas for different uses within a ground plan. > see chapter Basics, Areas for different uses Thus, for example, noise from busy roads, or being too overlooked by neighbouring buildings, can be prevented at the design stage by using ancillary rooms to screen off sensitive areas or by minimizing façade apertures appropriately. In contrast, particular views and quiet or protected outdoor areas offer the possibility of opening up the façade deliberately to include the outside space in the design for the interior.

Use times

The principal use times and fundamental lighting requirements of the various usable areas in a home can be determined by relating the rooms to the path of the sun. > see Fig. 1 In this way, the orientation of the building can create areas that will be lit in different ways, and that will take account of the various demands on utilization. > see chapter Basics, Areas for different uses

> 🛈
Points of the compass

A clear east-west or north-south orientation for a building is favourable in this context, but intermediate solutions may produce good results if the rooms are oriented carefully.

The north side is characterized by little sun and even light. Planners can place the entrance here, or storage and ancillary spaces, as they generally need little light. The sun rises in the east, and when it is low in the sky it will provide sunlight for areas that are used particularly in the morning. This is a good position for kitchens, adults' bedrooms and bathrooms. The south side receives the greatest proportion of sunlight. Areas for children and dining, terraces and conservatories, and other areas that are used particularly from the late morning through to the afternoon benefit most. The sun sets in the west, and can light areas that

🛈
\\ Hint:
For an observer in the northern hemisphere, all heavenly bodies rise in the east, reach their highest point in the south and set in the west. In the southern hemisphere, these conditions are reversed: the bodies still rise in the east and set in the west, but reach their highest point in the north. Thus, the orientation information applies only to the northern hemisphere, and must be switched for the southern hemisphere.

13

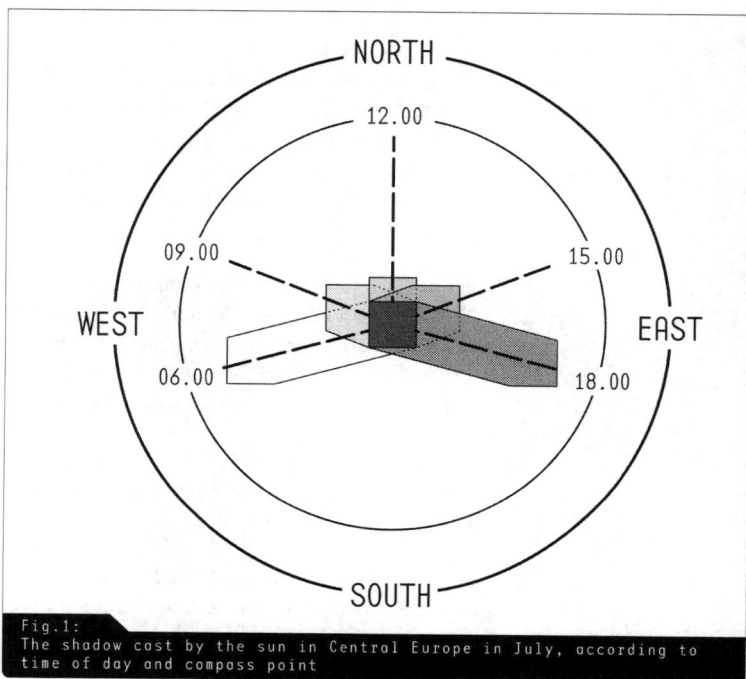

NORTH

12.00

09.00   15.00

WEST   EAST

06.00   18.00

SOUTH

Fig.1:
The shadow cast by the sun in Central Europe in July, according to
time of day and compass point

are used particularly in the afternoon and evening, such as general living and leisure areas.

AREAS FOR DIFFERENT USES

A home allocates use areas for everyday events, and these structure the ground plan.

Living areas   Living areas should be equipped so that they are appealing places in which to spend time. The way they are used is greatly influenced by individual needs and the rhythm of the occupants' lives. The needs include living rooms and bedrooms, and comparable areas in a home that can be used individually. › see chapter Elements of living

Functional areas   Functional areas such as the kitchen, bathroom and, where applicable, special working areas, already have their use fixed. They need a

particular infrastructure for water supply and drainage, and their use can only be changed by elaborate interventions into the structure of the building. It is the functional areas that first make a set of rooms into a dwelling, and enable the occupants to be largely independent of the outside world.
〉 see chapter Elements of living

Circulation
areas
Internal circulation areas, such as corridors, halls and stairs, separate and connect the different areas of a home. The arrangement and design of the circulation areas considerably affects the quality of living, as they determine the sequence of rooms in a home and can impose a hierarchy on transitions. According to the design approach, they may have spatial qualities in addition to their access function, and be used temporarily or permanently. 〉 see chapter Elements of living, Circulation areas

ZONING
The areas for different use in a home described above – the living, function and circulation areas – are interdependent and interwoven within a use matrix. A dwelling can be structured into zones in which different use areas join to form a unit. Thus, for example, intimate areas such as bedrooms and bathroom, possibly combined with an anteroom, can be combined to create such a unit. 〉 see Fig. 2 The spatial structure of kitchen, dining and living room can also join to produce a zone that has a more general character within the home. 〉 see Fig. 3

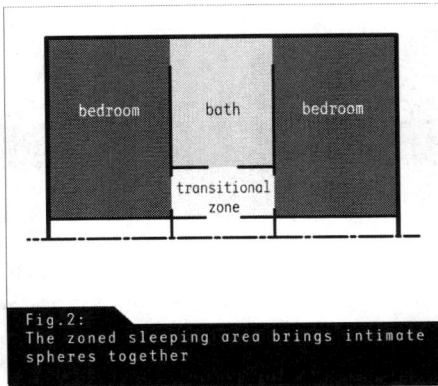

Fig.2:
The zoned sleeping area brings intimate spheres together

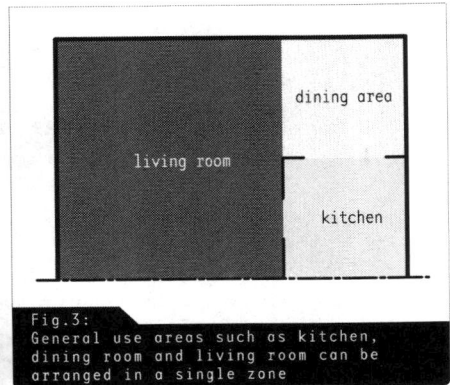

Fig.3:
General use areas such as kitchen, dining room and living room can be arranged in a single zone

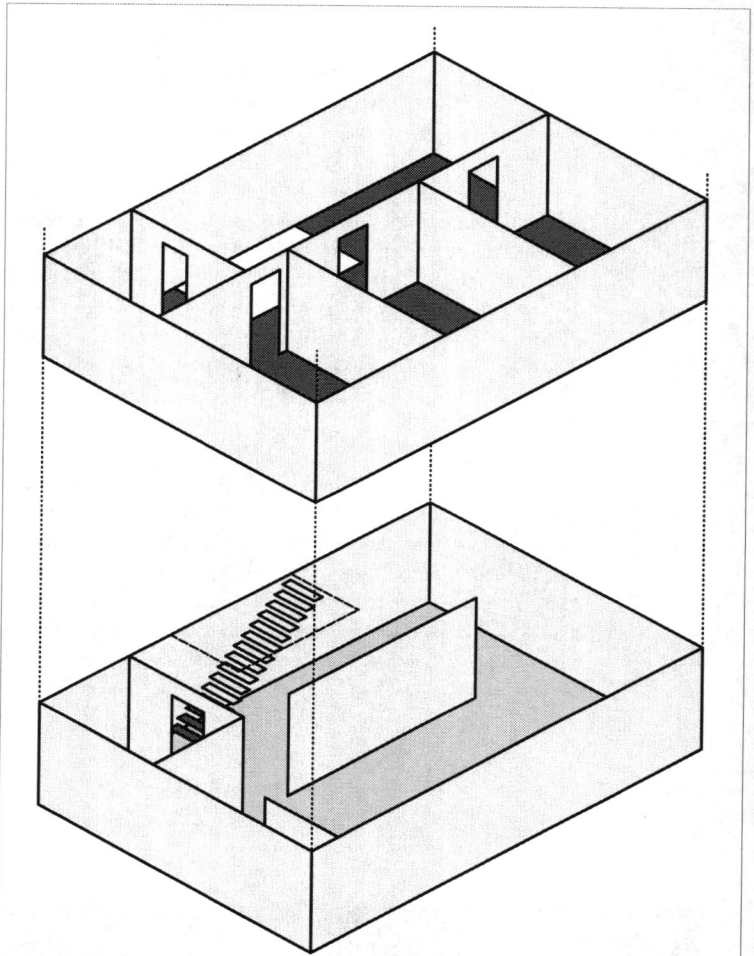

Fig.4:
A living concept with more than one storey allocates different uses
to the floors; the intermediate floor can shelter them

If a great deal of space and very different use areas are needed, designing a ground plan on one level can be difficult, as constraints in access and lighting emerge, and the building outline cannot often be freely chosen. A design for living on more than one storey can structure the various use areas on horizontal levels, and thus divide them according to their

Table 1: Specimen criteria for zoning a home		
General area	↔	Private area
Working area	↔	Leisure area
Living area	↔	Function area
Circulation area	↔	Living area
Adults	↔	Children
Link with outdoors	↔	Link with indoors
Vertical links	↔	Horizontal links
Extroverted	↔	Introverted
Open	↔	Closed
Noisy	↔	Quiet
Light	↔	Dark
Day	↔	Night

different uses. > see Fig. 4 For example, an entrance level offers space for more general uses, while rooms for private use are placed on a different floor, and thus screened off spatially by the intermediate floor. Single-family homes, as well as apartments in multi-storey houses, can include two or more floors, allocated to the different use areas. > see chapters Basics, Creating space, and Building forms

Criteria

Possible criteria for zoning a dwelling can also be periods of use, the need for a special work area, and thematic links defined by the occupant's needs. Allocation to the various categories can help to order the related quality or contrasts in use demands within the home. Thus, qualities like noisy/quiet, open/closed and introverted/extroverted are used to develop connections and match them to each other. > see Table 1

CREATING SPACE

The architectural design creates areas and volumes. Adding in the different use areas creates two- and three-dimensional connections, permitting visual links, perspective spatial impressions and communication possibilities. The rooms in a home are defined by ground plan form and volume. Length, width and height can be matched to a number of individual needs and spatial relationships. A particular room can be allocated a specially adapted area and an appropriate volume. In contrast, rooms can be treated use-neutrally and with equal value, so that differences in size do not impose a hierarchy. Another design approach can divide a coherent area or a complete volume into different areas. These different design approaches can be applied throughout a home or can characterize different areas.

> $\mathcal{P}$

Flexibility

In the context of housing construction, flexibility describes the possibility of making individual rooms or a whole residential unit adaptable. The above-mentioned use-neutral room concepts and the open ground plans described below offer starting points for being flexible when changing and adapting rooms while they are in use.

> ⓪

Purpose-built rooms

If rooms are intended only for a designated use, they are purpose-built. Here, the area, scale and connections with other rooms are fixed under consideration of use-specific qualities. > see Fig. 5 The following specifications are customary in detached house building: the living room is usually the largest room, followed by the parents' bedroom, children's bedrooms and the functional areas including kitchen, hygiene and ancillary rooms. In a case like this it is not easy to reallocate room uses if the occupants' ideas change, and it is often impossible to divide the space up differently without a lot of expense. Normally the entire concept of a detached house is directed to the needs and wishes of people, with no long-term change to the home's structure intended.

Use-neutral rooms

If rooms are largely identical in their situation, size and shape, they are use-neutral. This approach makes sense for rented accommodation, for example, which is definitely intended to be used by different occupants over the course of time, as it is not possible to anticipate all the needs of future occupants or changing uses. The rooms are not structured hierarchically, and are interchangeable in terms of use. > see Fig. 6 One of the things this approach explains is the popularity of refurbished homes dating from

$\mathcal{P}$

\\ Example:
The architect Adolf Loos's "Raumplan" (Spatial plan) concept deals with various rooms strictly according to use, area and volume. The ground plan shows a marked independence among the rooms, but as an overall concept them seem like a coherent spatial structure; Müller House in Prague, 1930.

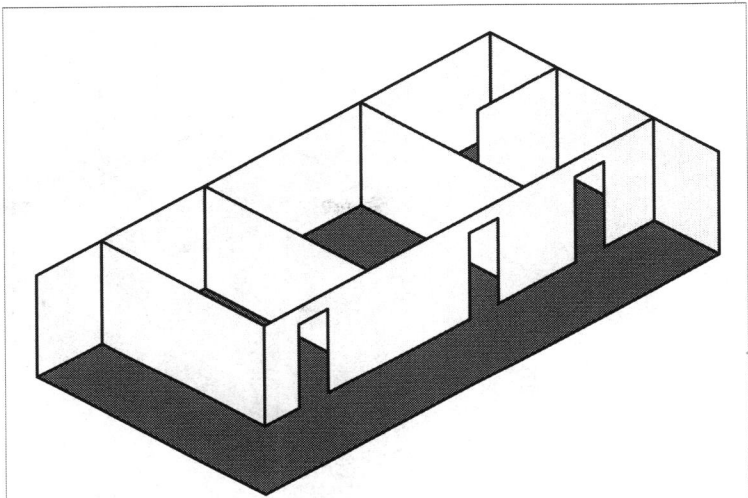

Fig.5:
The form of a purpose-built room is largely adapted to a special use

the early 1870s in Europe, which often have use-neutral ground plans. So if necessary the internal arrangement relating to use can be redefined largely without spatial constraints. It is not just individuals and families who benefit from a home structure of this kind. It is now not just students, but also increasingly professional and elderly people, who share houses and therefore like this kind of flexibility.

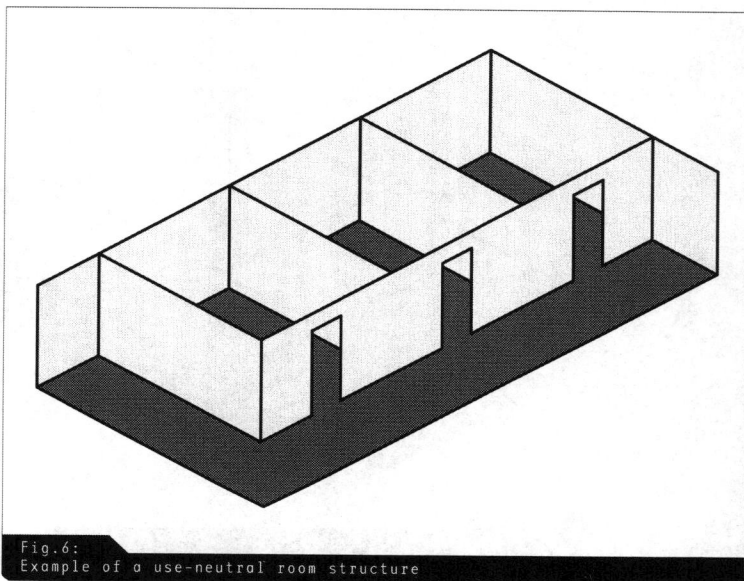

Fig.6:
Example of a use-neutral room structure

Open ground
plans

Open ground plans combine various uses within a home, within a set of fluent transitions. This design approach creates generously coherent spatial volumes that are not divided by intricate access structures and intermediate walls. Smaller houses and apartments benefit, not least because space is not used for access alone. The clearly delimited spaces created by other dwelling forms are replaced by zoning in an open ground plan, and the zones can be accentuated by changes of material or colour, or of natural light and artificial illumination. › see chapter Basics, Zoning In addition, the open ground plan can be differentiated individually using temporary or (semi-)closed room dividers in the form of open shelf units, (transparent) sliding doors and screens, and thus allow changing spatial impressions and sightlines.

The bathroom and WC are the only functional areas that require a minimum of fixing and isolation, but particular design approaches can blur the boundaries here as well. For example, they can be brought together in a core and positioned centrally in the ground plan. This arrangement produces structurally zoned spaces within a larger area, to accommodate different use areas. › see Fig. 7

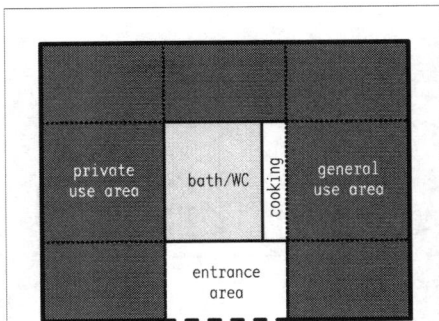

Fig.7:
Systematic arrangement of use areas around a functional core in an open ground plan

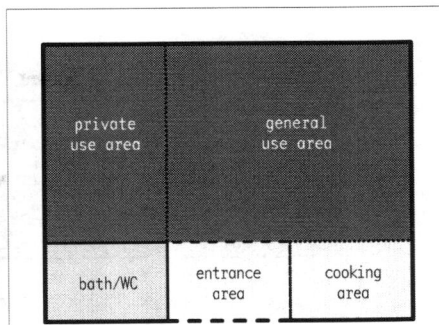

Fig.8:
Systematic combination and separation of use and functional areas in an open ground plan

In addition, private and general living areas can be combined with a kitchen and dining area to give a manageable and generous impression of space. However, in this case an entrance area should be created as a transition from outdoors to indoors, so that the living area is not entered immediately from the outside. > see Fig. 8

Free division makes the open ground plan very flexible for use, and it can deal with many different demands and needs. The single-person household, which is occurring ever more frequently, as well as couples and, temporarily, young families, find this structure appealing. An open ground plan is suitable only to a limited extent for a larger number of occupants with different requirements, as noise and overlapping uses can be disturbing.

Proportions and room height

The height of a room is perceived subjectively and is linked with its ground area. The more open and larger a room is in area, the lower its height will seem to be. > see Fig. 9

Selected use areas can be enhanced by greater room height, especially if they are to be emphasized within the general living concept. So it is possible to accentuate individually important areas, e.g. the living room or dining area, by means of special spatial qualities, and to design other areas of the home more reticently and less voluminously where necessary. > see Fig. 10

21

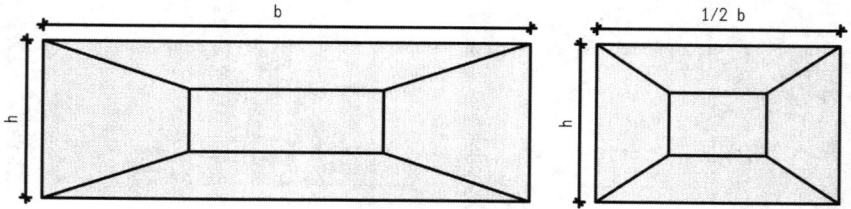

**Different
levels in a
room volume**

Rooms may also be high enough to be divided into two levels within a single volume. For example, a sleeping area can be fitted with a gallery, or a higher retreat zone can be created in a generally used living area. › see Fig. 11

Planners do not always have the design freedom to choose storey heights for homes in terms of spatial considerations alone, as larger volumes make for additional construction and heating costs. Restrictions to the overall height of a building, in multi-storey housing construction, for example, can restrict design freedom with respect to the room height of individual floors.

**Homes with more
than one floor**

In residential units with more than one floor, two or more floors can be combined spatially as well as functionally to create special and changing spatial impressions. The necessary vertical access is then differentiated by means of open or closed stairs, which can downplay or intensify the horizontal separation of the floors. It is not just the necessary staircases that offer possible connections: air spaces can present spatial volumes in parts of selected areas. Galleries, often linked to the lower storey by additional stairs, permit direct communication between the floors. › see Fig. 12

Maisonettes

In multi-storey housing construction, the forms produced are called maisonettes – a little house within a larger housing complex. › see Fig. 13 Such homes in particular can be given special spatial qualities by galleries and air spaces.

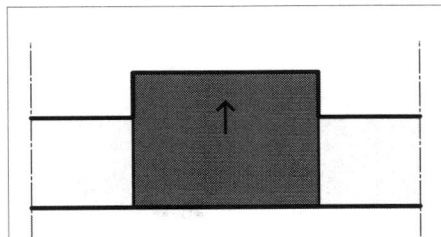

Fig.10:
Particular areas of a home are
emphasized by greater room height

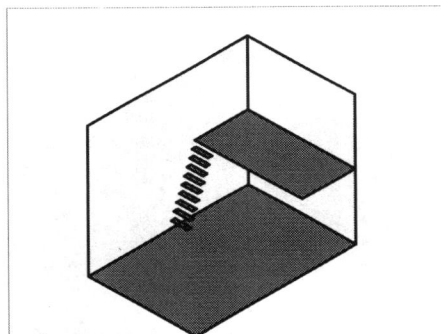

Fig.11:
A gallery creates a second level within
a room volume

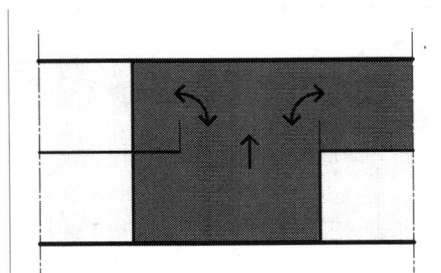

Fig.12:
Air spaces enhance particular areas
and can create direct communication and
sightlines between two floors

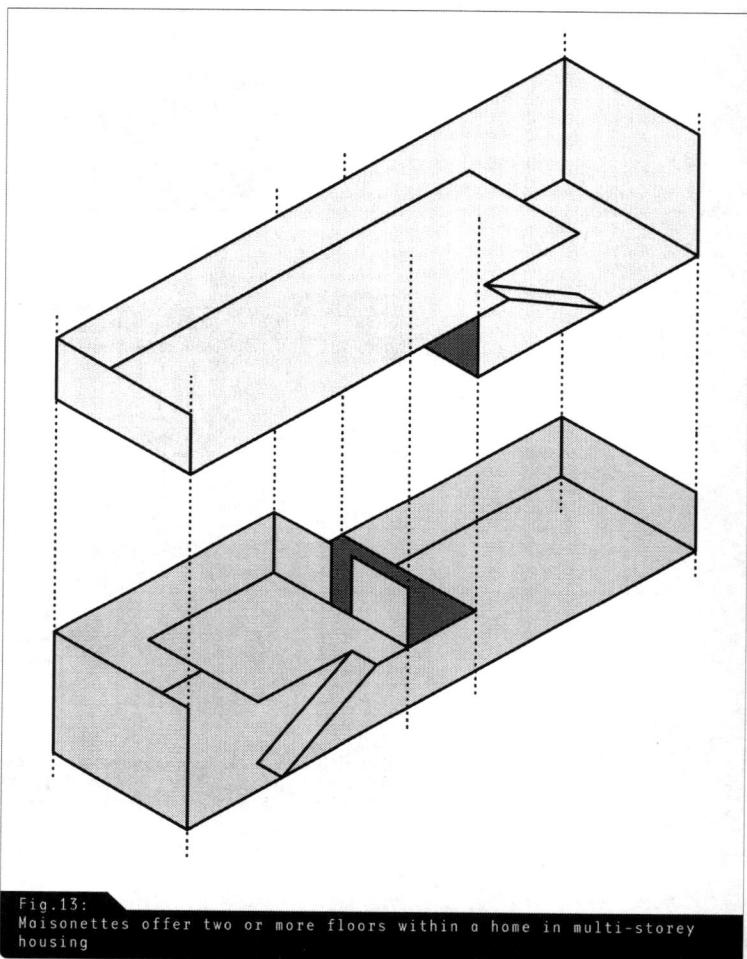

Split-level homes

Split-level homes are a particular form of the maisonette. > see Fig. 14 Here, life is conducted on various levels, occupying only one part of the building depth available, and linked by internal stairs. This arrangement means that even comparatively small homes can be zoned on various levels. > see chapter Basics, Zoning

Linking use areas

The transitions from one use area to another may have different design characteristics and thus create spatial accents. Fluent transitions

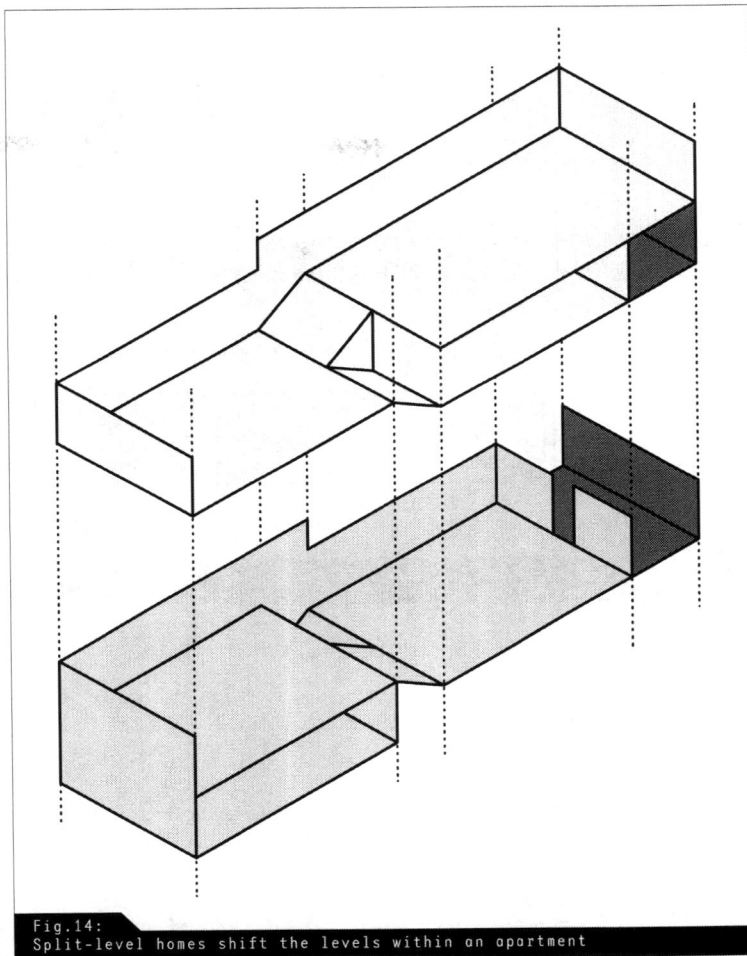

Fig.14:
Split-level homes shift the levels within an apartment

without fixed limits permit largely open perspectives and create a coherent spatial concept. In contrast, structures that form walls divide the various uses and define the different areas clearly. The connecting apertures between the areas influence the effects of the changing space. The width of an aperture is just as important in this context as its height. For example, rather than have a relatively low door, it is possible to create a floor-to-ceiling opening in a wall. A narrow aperture also makes a different effect from a lavish gateway. The individual access points impose a hierarchy on

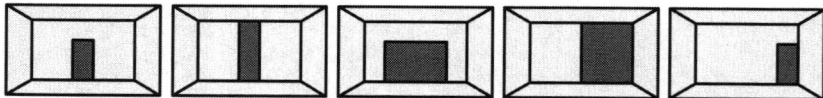

Fig.15:
Different door apertures affect the change of space and impose a hierarchy of access types

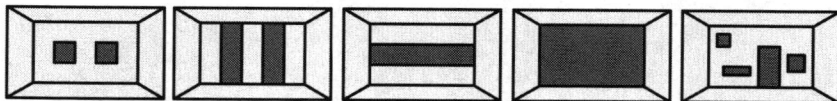

Fig.16:
Window apertures considerably affect links with the outdoor space and thus help to give the living space its character

the use areas and express introversion or extroversion. › see Fig. 15 Sliding and hinged doors make it possible to handle divisions or connections flexibly.

Wall apertures do not have to be used exclusively to provide access to use areas. It is possible to use perforations that also simply create visual connections with other rooms, or that can be used functionally as hatches. Circulation areas do not serve their purpose alone either, but can emphasize a change of area and define a spatial transition.

Links with
outdoors

Outdoor space is an important factor for the spatial concept as a whole in this context. Window apertures create visual links with particular local features, thus making their mark on the interior. › see Fig. 16 The perspective is changed by an increase in the height level of a storey. A ground-floor apartment may have the advantage of access at ground level and/or a connection with the garden, but it can also mean that the home is undesirably overlooked from the outside. Measures should be taken where necessary to ensure privacy. The higher the floor level of a home is, the further the eye can roam, and the living space is all the more protected from being overlooked.

›✏

26

Exits, balconies and terraces combine outdoor and indoor areas; outside areas can become part of the living space. Such a link can be emphasized by space-defining structures in the form of extended walls and pergolas, as well as the use of uniform materials.

ELEMENTS OF LIVING

The living, function and circulation areas mentioned above are key elements of housing construction design. › see chapter Basics, Areas for different uses These areas can be differentiated with regard to individual needs and general requirements to form the different elements of a dwelling.

SLEEPING

Sleeping is a basic human need. The quality of one's surroundings affects sleep significantly. This special environment must take very different requirements into account and can be characterized by various conceptual approaches. The bedroom area is principally devoted to the individual phases of rest and recovery, and can be defined by considering these criteria exclusively. This means that the bedroom area is monofunctional and clearly separated from other use areas.

But it is also possible to create potential over and above the principal use and allocate other uses within the sleeping area. It then becomes a general location for the intimate sphere and private purposes according to the time of day.

Various access links are created if other necessary use areas are added. › see Fig. 17 If a dressing room is to be provided, it is an advantage for it to be linked directly with the bedroom, as privacy is destroyed by having to pass through an intermediate corridor. Walk-in wardrobes can be a space-saving and sensible alternative. It also makes sense to be close to the bathroom, to avoid passing through areas that may be used communally. Ideally, direct access should be planned, without a detour via an intermediate corridor.

For reasons of noise insulation, the sleeping area should be screened from rooms that are used communally. If horizontal zoning on different floors is not possible, direct access through "noisy" areas can be avoided by adding in transitional zones. › see chapter Basics, Zoning This recommendation does not necessarily apply to a single-person household, as the uses here are tailored to the daily rhythms of one individual, and internal disturbances by other people are not usually anticipated.

Sleeping areas can be matched to individual needs by various design approaches. A bedroom that is almost closed off from the inside and outside worlds, and used exclusively for sleeping, can thus represent the most

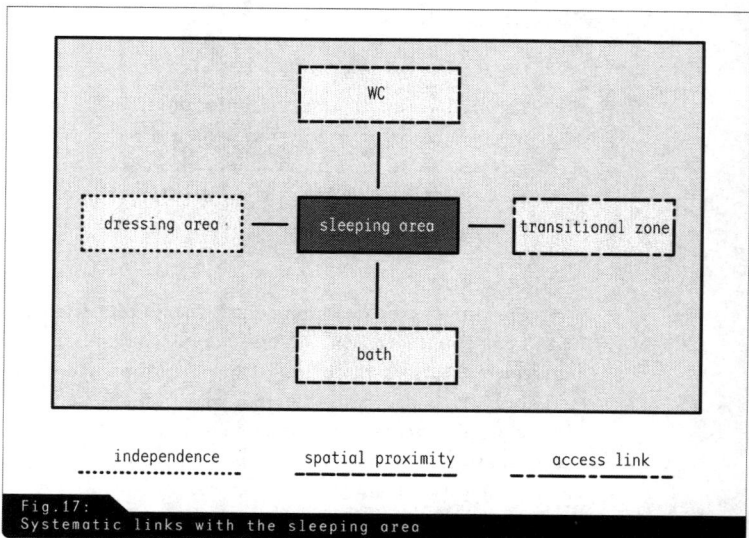

```
                    ┌─────────────┐
                    │     WC      │
                    └─────────────┘
                          │
┌──────────────┐   ┌─────────────┐   ┌──────────────────┐
│ dressing area│───│ sleeping area│───│ transitional zone│
└──────────────┘   └─────────────┘   └──────────────────┘
                          │
                    ┌─────────────┐
                    │    bath     │
                    └─────────────┘
```

independence   spatial proximity   access link
······················· ───────────────── ─ ─ ─ ─ ─ ─ ─

Fig.17:
Systematic links with the sleeping area

extreme form of introverted sleeping area. › see Fig. 18 left External influences that could disturb restful sleep are thus excluded in the planning. Reticent access apertures can underline the room's function as a screened-off area of retreat. Variations in size and proportions create an impression of generous space, or even ascetically functional aspects, if they – as a radical example – are modelled on a ship's bunk.

Sleeping areas relating to the outside

Including the outside area can downplay this introverted quality. A well-placed window or large glazed area, taking orientation into account, changes the light situation considerably and can create special spatial conditions. › see chapter Basics, Orientation In this way, window apertures can provide a source of indirect light and scarcely offer any view out at all. Special window shapes can frame a perspective on the outside world if the viewer is in a particular position, such as lying on the bed. Taking the idea further, in a suitably private situation, the whole outer façade of the bedroom area can be made up of glass elements, which largely cancels out the border between inside and outside. › see Fig. 18 centre left and centre right

Sleeping areas with integrative uses

The bedrooms can also be designed more openly to accommodate other function and use areas, according to access and requirements. For

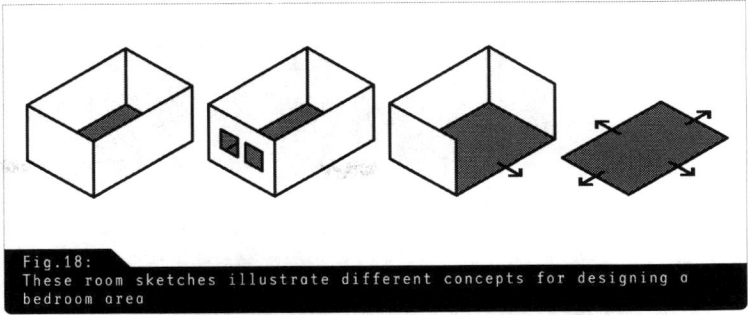

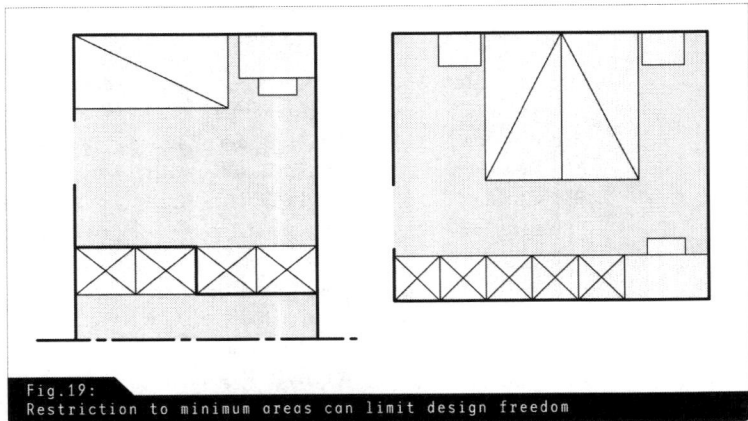

example, individual areas used for work or leisure can be combined with the sleeping area because these activities take place at different times. › see Fig. 18 centre right

In its most extroverted form, the sleeping area can be placed within an open ground plan and be little different from the other functions and uses in the home. › see Fig. 18 right, and chapter Creating space

Furnishing   Like every use area, the bedroom fundamentally needs sufficient space for furniture, fittings and circulation. › see Fig. 19

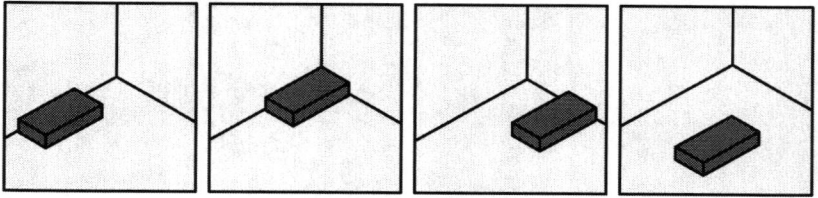

Fig.20:
The position of a bed in a room affects the sense of security and rest of the person lying in it

The smaller the dimensions of these spaces, the more alternative design is limited. Optimizing areas is certainly a desirable planning feature, but it should impede flexibility only to a limited extent, and should permit variable design approaches in terms of subsequent use as well. For example, there are standard-size beds that define how much space is needed. But intermediate sizes, self-built variants, four-poster and raised beds are all places to sleep that do not have to come in standard sizes and may be part of an individual sleeping area.

> Positioning the bed

Positioning the bed in a room affects conscious and subconscious perceptions. If the bed is in a corner, the whole room and its openings can be in sight, suggesting clarity and security. If the bed is placed centrally, and is also given focus by the position of room apertures and the position of other furnishings, it can be a special centre. > see Fig. 20

\\ Hint:
The length of an individual bed can be worked out by adding approx. 25 cm to the height of the person concerned. Room to move around the bed should be planned to guarantee accessibility. Generally speaking, areas for one person to move in should not be less than 70 cm.

The sleeping area is often also the place where clothes are changed, and therefore needs working space for dressing and the items of clothing involved. > see chapter Elements of living, Storage If there are no plans for a separate dressing room, space should be left for a wardrobe. Care should be taken that there is enough room to move around it, and that it does not stand side on to the door axis and thus hinder access to the room. Built-in floor-to-ceiling cupboards may be an alternative, as they use the space available in the best possible way. However, this does tie up areas and sections of wall, and could possibly constrain other, later design approaches.

EATING

Few social activities have been cultivated traditionally to such an extent as eating. It is celebrated at political and cultural events, typical local dishes define regional and national self-awareness, and special events culminate in a festive meal at home or in a restaurant with our circle of family and friends. It is all the more remarkable that little time is often allowed for everyday breakfasts, lunches or dinners, as work and active leisure are given priority. The status of an eating area in a home thus depends on individual requirements. The number of occupants is important, as well as individual eating habits, and the space available. The eating area also plays a part beyond its actual function as a daily meeting point for the occupants, and for social gatherings. Compact, functional or spacious eating areas are thus designed to suit demands and requirements.

The eating area is usually served by the kitchen and should therefore be placed close to it. Long distances make it more difficult to present food and clear dirty dishes away. If the eating area is outside the kitchen, the working processes can be optimized by a hatch with a space to put things on in front of it. Proximity to the front door and cloakroom should also be taken into account, so that guests can find they way to the laid table or the WC without detours. > see Fig. 21

A clearly delineated space can be allocated to the eating area, thus emphasizing its significance within the home, and its independent and self-contained function. If the room is also to be used for larger gatherings of family or friends, for example, the appropriate space must be planned. Alternatively, the eating area can be placed in a general living area in an all-embracing open living concept. > see chapter Basics, Creating space A constellation of this kind draws no definite borders, and combining areas and functions creates extensive visual links and an impression of generous space. Another advantage is flexible use, as the table can be extended if necessary without constraints on space. If the eating area is to be emphasized in an

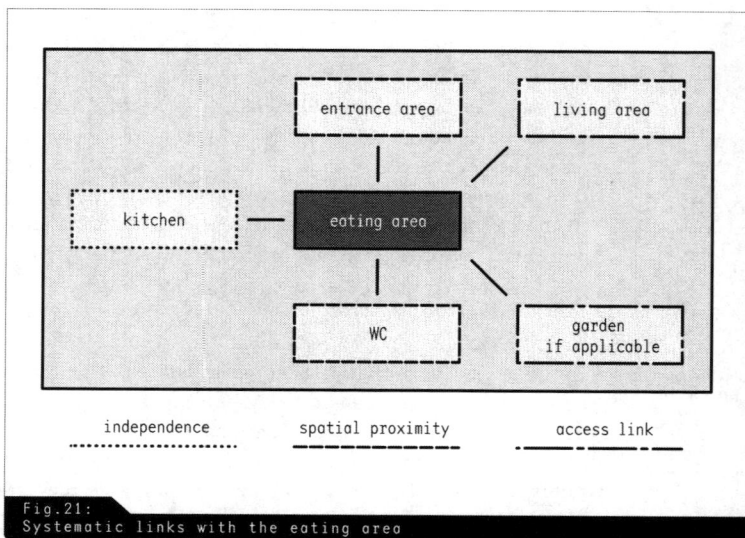

```
            ┌─────────────────┐          ┌─────────────────┐
            │  entrance area  │          │   living area   │
            └─────────────────┘          └─────────────────┘
                     │                  ╱
  ┌─────────────┐        ┌─────────────────┐
  │   kitchen   │────────│   eating area   │
  └─────────────┘        └─────────────────┘
                     │                  ╲
            ┌─────────────────┐          ┌─────────────────┐
            │       WC        │          │     garden      │
            └─────────────────┘          │  if applicable  │
                                         └─────────────────┘
```

independence   spatial proximity   access link

Fig. 21:
Systematic links with the eating area

open spatial situation, the obvious thing to do would be to pick up links within the space, and to use varieties of material and lighting creatively.

Another possibility is to integrate a dining area within a kitchen that also serves as part of the living area. › see chapter **Elements of living, Cooking** This approach lessens distances to be covered and also enables communication between areas for eating and for preparing food. It is also conceivable in principle to have two eating areas in a home, making a distinction between a functional area for snacks and breakfast in the kitchen and another area elsewhere for the main meals. › see Fig. 22

Eating in the kitchen

Links with the space outside a home can fundamentally influence the quality of the eating area. Windows coming down to ground level, exits or (roof) terraces can make the outdoor space a conceptual part of the indoor one. The light available and the position in relation to the sun also affect the room, and will help to shape the design, taking the different times of day into consideration. › see chapter **Basics, Orientation**

Links with the outside area

The arrangement of the furniture and the amount of space needed for furnishing an eating area depend on the spatial concept and the number of users.

Furniture

›

34

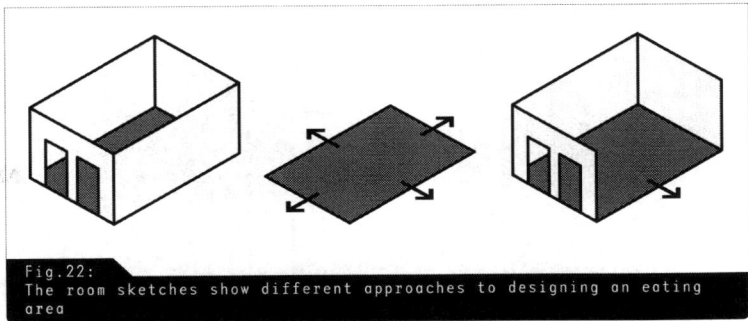

Dining tables can be placed in various different ways and in a variety of spatial situations. Round tables make it possible for places to be laid equally, while rectangular tables demand a directed arrangement. A freestanding table becomes the dominant element and establishes the key links in the spatial structure. In a corner situation, possibly combined with a corner bench, the eating area will look more essentially functional. Work surfaces can also be used temporarily for a compact snack area inside a kitchen. > see Fig. 23

Uses and
functions

WORKING

Work has an important part to play in our everyday life. It determines our rhythm of life, our environment and our social position in society. Our home situation is also influenced by work activity – materially, because

> \\ Hint:
> One person needs a space at table at least 40 cm deep and 65 cm wide to be able to eat without getting in the way of neighbours. A rectangular dining table for six people will be at least approx. 80–90 cm wide and approx. 180–200 cm long, and the round variant will have a diameter of approx. 125 cm (formula: place width in cm x number of people / 314). If the table is freestanding in the room, the clearance between it and the nearest wall should be at least 1 m, so that it is possible to pass people who are already sitting down.

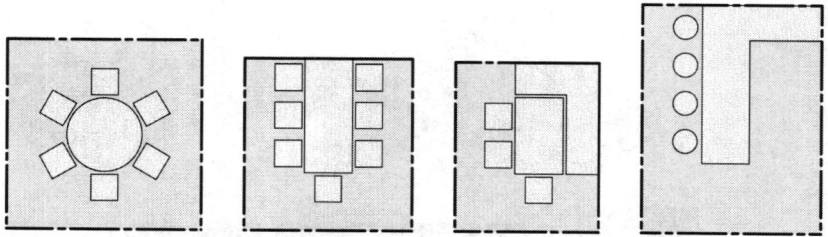

it often determines the amount of money available, but also functionally, as various work activities make special demands on a home. Almost every home offers some facilities for brainwork. Communications and computer technology in particularly, with mobile phones, laptop and the Internet, make it possible to work almost anywhere. Thus the bedroom, the dining table or the living-room sofa become temporary workplaces.

But it is often not possible to guarantee an ergonomic, long-term workplace in locations of this kind for various reasons. Lighting conditions, noises and overlapping uses in particular may necessitate having a special home working area if work is to be done regularly.

Room concepts

Designing a working area and creating a sensible categorization and separation in the home depend on requirements, and these are defined by the work involved. Classical office activities like book-keeping, for example, can usually be pursued in a small space, provided that it is designed functionally. For example, if a working area of this kind is placed near the entrance to the home, it is possible to create an independent area for receiving work-related visitors without involving private areas of the home. But as the need arises, the working area may also run into other use areas in an open ground plan. > see chapter Basics, Creating space A number of professions can be pursued at home, shaping and differentiating it through the activity's particular demands. For example, a painter's studio, a sculptor's workshop, a sound studio or a musician's practice room, or even a teacher's study create particular emphases that can scarcely be detached from other use areas and also permit interaction between work and leisure. Here, zoning can suggest helpful approaches for achieving good solutions. ❭ see chapter Basics, Zoning Any links with the outside world that might be needed, lighting and other special criteria such as sound insulation depend on the particular demands of the activity pursued.

RECREATION AND LEISURE

Uses and functions

Many leisure activities take place in an enclosed space, and space is generally provided for recreation and relaxation in the home. A bed is not used exclusively for sleeping, the dining table can also be used for games, the bathtub can be a place for relaxation, and a living room with television and reading area is almost taken for granted as a component of many homes. Often very little intervention is needed to make every room in the home a place with potential for recreation and leisure.

Room concepts

The living area can be central to life in the home and to eating, and is often also used for recreation in the course of everyday leisure. ❭ see Fig. 24

The living room is usually an extrovert part of the home, a central area for spending time, the qualities of which can be enjoyed by guests as well as the occupants. But there are also introverted room concepts that can make the living room into a private area of the home, screened-off inside and outside.

A large living area conveys a generous impression of space and can also be subdivided, for example into reading and play zones, and spaces for communication and multimedia use. ❭ see Fig. 25

A recreation area can be designed with special spatial qualities, e.g. with greater height, or in some cases linked with other uses through air spaces and galleries. ❭ see chapter Basics, Creating space Such expansive planning approaches cannot always be realized, and small and compact living spaces can also be handled in such a way as to create a high-quality area in which to spend time, particularly emphasizing the manageable and comfortable aspects of living. There is also the possibility of adding spaces for other main uses, e.g. eating area and kitchen, to a (small) living room and thus implementing the spatial concept of an open ground plan with various zones. ❭ see chapter Basics, Zoning

Fitness areas, a music room, a library, or other individual leisure areas leave a great deal of scope in the design for creative and integrative solutions. Whether the use is planned to be temporary or long-term will

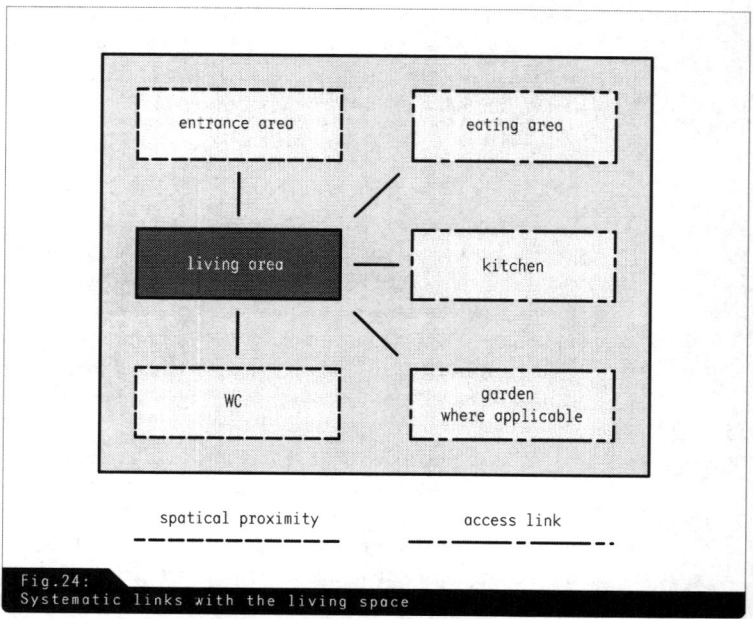

entrance area

eating area

living area

kitchen

WC

garden
where applicable

spatical proximity   access link

Fig.24:
Systematic links with the living space

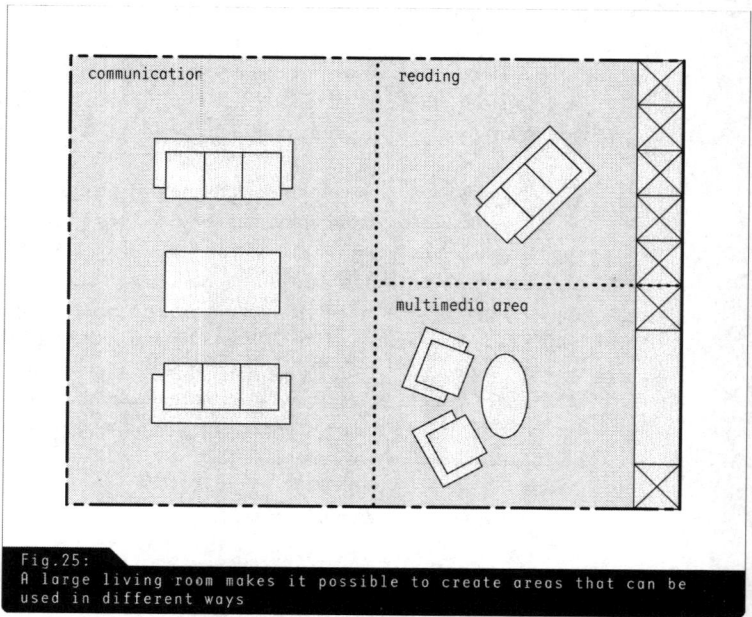

communication

reading

multimedia area

Fig.25:
A large living room makes it possible to create areas that can be
used in different ways

38

be defined by the occupant's personal requirements. In contrast, there are areas dedicated exclusively to recreation or special leisure activities, such as saunas and private swimming pools. Special children's play areas need possible visual links with other use areas, so that adults can keep in personal contact and supervise the children, ideally at the same time as doing other things. Open kitchen-dining rooms, living areas, and passageways offer a good opportunity in this sense: an eye can be kept on them from any adjacent area. > see chapter Elements of living, Circulation areas

> 0

Links with the
outdoor space

Recreation and leisure areas inside the home benefit greatly from including outdoor space. For example, ground floor apartments can build outdoor spaces into the interior concept if the weather is suitable, and thus extend the living space. > see Fig. 26 Large or small balconies and roof terraces can meet similar criteria on the upper floors. As mentioned earlier, the light and position in relation to the sun are important: it is recommended that generous apertures be created for the afternoon and evening sun, so that the living and leisure area can be naturally lit at the principal times the space is used.

HYGIENE

Uses and
functions

Physical hygiene is expressed in various forms. Whatever individual cleaning rituals may look like, they are important to life and health. Water is the central theme in hygiene, and traditionally plays a major part in many cultures. Seen entirely practically, the water has to be supplied

0

\\ Hint:
Little extras for children can be built into
a home without a great deal of effort. To
achieve this, an attempt can be made to see
a design for a home or a well-built house
through the eyes of a child. Small niches
become potential hiding places, the table
that folds up against the wall can become an
Indian tepee with the help of a tablecloth,
and a small corner window mysteriously lets
the sun shine on a tile pattern at a certain
time. Children are masters of design and
improvisation. Everything they need is in
their imagination and in tiny elements of room
design.

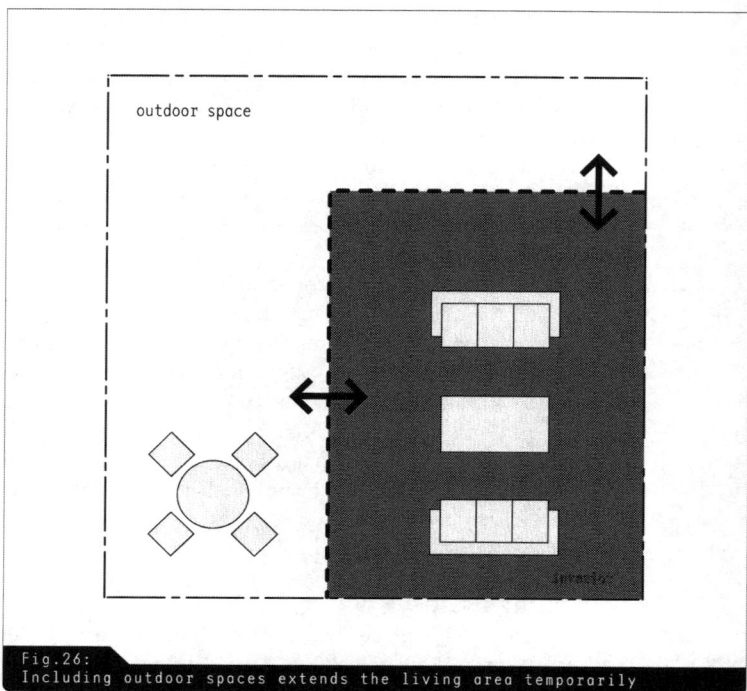

in a completely pure condition and disposed of after use. The necessary infrastructure makes a bathroom a very specialized function area within the home. Unlike other areas it cannot be "converted" without difficulty, and should address the needs of different users. The bathroom and guest WC need a certain among of sight and sound insulation, so as not to affect other use areas, or be affected by them.

It makes sense for bath and private WC to be placed in the immediate vicinity of the sleeping area, so that there can be direct access to them. If this is impossible, access to the bathroom, as well as to the bedroom area, should be screened from eye contact from less private areas. The cloakroom is placed in the entrance area of a home where possible, and can serve the general living areas here. > see Fig. 27

If a home ground plan is restricted in size, several functions are often grouped together in a bathroom. The WC can be in here as well as the

🔖

\\ Hint:
When planning, particularly for housing with
more than one storey, the walls in which
plumbing for fresh and sewage water are
fitted should be one above the other wherever
possible, to reduce installation difficulties
and noise pollution. Good sound insulation
should be considered at the planning stage,
especially for adjacent homes. For example,
the plumbing walls in two adjacent homes can
be next to each other and thus not affect any
other rooms. To keep the pipe runs short, the
sanitary equipment should be fixed as close to
the plumbing walls as possible.

bathtub, shower and washbasin, perhaps with bidet, baby-changing table, washing machine and tumble dryer. Bathrooms are all too commonly rooms with minimal space and a high degree of functionalization. The fundamental question arises of whether things have been combined here that actually belong together. In the above example, three areas overlap: personal hygiene, washing clothes and using the WC. Thought could be given to whether the standard of hygiene in many ground plan solutions in old and new buildings do justice to functions and users. If an area for ancillary functions is planned, the laundry area could be placed elsewhere.
> see chapter Elements of living, Ancillary functions Alternatively, a washing machine can also be placed in the kitchen, in the dressing area or in other ancillary rooms. If use of the WC and the personal hygiene area can also be separated spatially, the bathroom can become a pleasant and clean place for relaxation and cleansing.

Room concepts   Strict application of such standards, however, seldom produces high-quality rooms. > see Fig. 28 The question of whether a space-saving bathroom with shower, washbasin and toilet should be planned, or whether it is possible to design a generous bathroom with a generous amount of space depends above all on the area available.

In order to create functional focal points, dividing elements, for example, can be installed in the bathroom to separate off special areas and also increase efficiency. > see Fig. 29 A compact personal hygiene area can be created with shower and washbasin, with a separate area for the bathtub. The toilet and the bidet, if fitted, are also separated off in the bathroom or

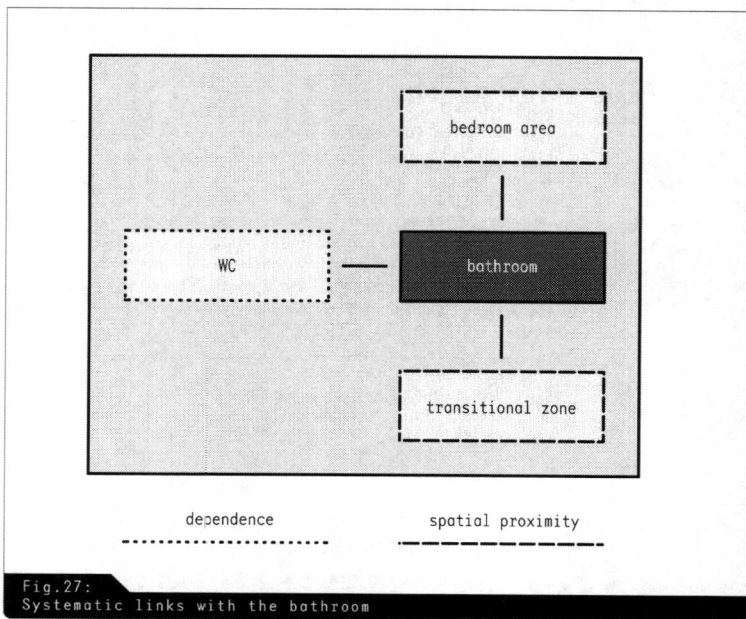

Fig.27:
Systematic links with the bathroom

allotted a rooms of their own. An arrangement of this kind can avoid block-ages and unpleasant odours, especially in the morning. In larger house-holds an additional WC (cloakroom) makes sense anyway.

Apart from its functional aspects, the bathroom can be seen as a place where it is pleasant to spend time. For example, it is possible to move away from the usual washbasin, bath and shower pattern and look for special ground plan and function solutions. A shower base can be re-placed by a custom-sized area with a floor-level drain. Or a (large) bath-

\\Hint:
Architectural visions for innovative and forward-looking bathrooms can be found in *Bathroom Unplugged* by Dirk Hebel and Jörg Stollmann (eds.), Birkhäuser Publishers, Basel 2005.

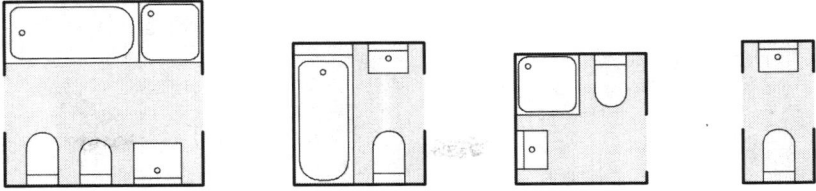

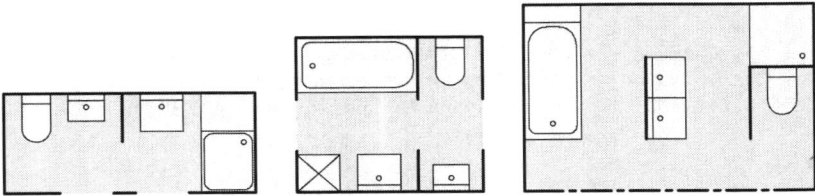

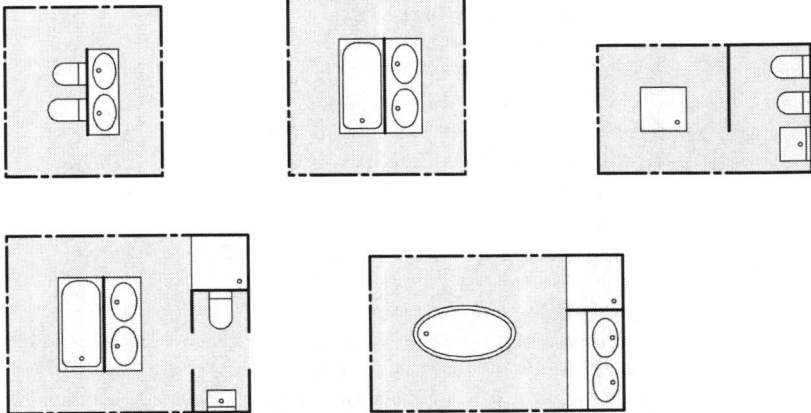

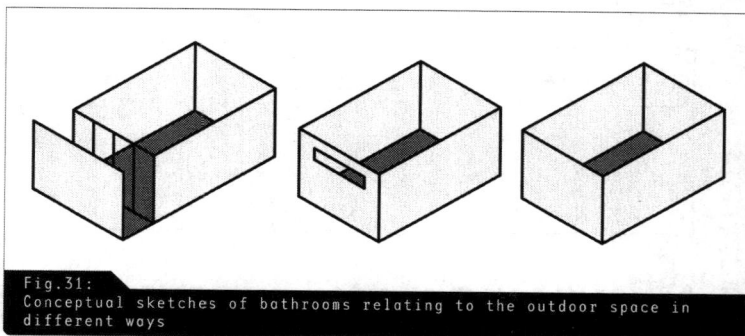

Fig.31:
Conceptual sketches of bathrooms relating to the outdoor space in different ways

tub is arranged standing free in the middle of the room, giving a sense of spaciousness and the extra room for moving about. Such solutions may well need more space, but enhance the quality of living in the bathroom function area. › see Fig. 30

Links with the
outside area Even though the bathroom is a very intimate area, its relationship with the outdoor space should not be neglected. If this outside area is not overlooked, a large area of glass or a small balcony can be very attractive. Cleverly chosen window apertures can create visual links with the outdoor space, without impairing the intimacy of the interior.

Bathrooms placed away from the outside walls do not allow any visual contact with the outside world or permit natural ventilation, and so have to be ventilated mechanically. On grounds of ventilation problems alone, windowless bathrooms are not as attractive to spend time in as baths with windows in the outside wall. One interesting variant for a bathroom without outside walls could be a glazed opening in the roof. This can take account of functional aspects of lighting and ventilation, and also gives a clear view of the sky from the bathtub. › see Fig. 31 right

COOKING

Uses and
functions Cooking has lost its vital importance in recent decades. It is no longer necessary for individuals to cook in order to eat. Fast food and ready meals determine a great deal of the menu for many Western people, and mass production means that the food industry can offer rock-bottom prices for tinned and frozen products. But cooking can be about more than preparing food in a way that is as time-saving as possible. It can be a communal

activity, carry a high level of leisure value and not least promote awareness of knowing what we eat.

As a rule the kitchen is a busy part of the home, and is thus of crucial importance. For example, it should not be too far away from the home's entrance so that food purchases do not have to be carried too far. Storage areas and vegetable gardens, where applicable, also need to be close to the kitchen. The eating area is served from the kitchen, and should be directly accessible, so that food can be put on the table and dirty dishes removed without undue difficulty. › see Fig. 32

Room concepts The kitchen can be defined in very different ways within a room concept. It can take up the space appropriate to a cooking niche, working kitchen or kitchen-dining room in its various forms, according to need and aspirations. If it is treated as an isolated functional unit it will be given a room of its own, clearly separated from other use areas. This room can perform its function through optimized areas and be merely a working area in which to cook. But it can also be made into an attractive place to spend time in and, for example, set up a snack area or a full eating area in a kitchen and thus create a central meeting place in the home. › see chapter Elements of living, Eating Alternatively, people often like to plan open kitchens, which will then merge into other use areas. › see

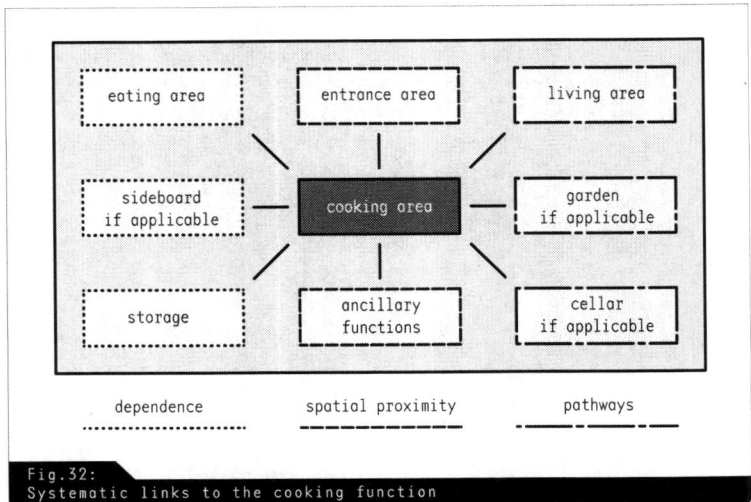

eating area	entrance area	living area
sideboard if applicable	cooking area	garden if applicable
storage	ancillary functions	cellar if applicable

dependence   spatial proximity   pathways

Fig. 32:
Systematic links to the cooking function

In this way, cooking as an activity is in communication with other living areas, and benefits from an integrated spatial concept. However, it should be noted that the open situation means that the cooking smells will affect the area for some time. In order to handle functional and spatial aspects flexibly, large (sliding) doors can provide temporary separation.

> ✎ , ◖

Links with outdoor space

Links with the space outside are essential for the kitchen, and not only for reasons of light and ventilation. If the kitchen is used frequently, by a family with children, for example, it is advantageous to be able to watch the children playing outside. In good weather, an attached terrace or balcony can be attractive for breakfast or another light meal.

> ◖ , ✎

Furnishings

The kitchen is divided up by work surfaces, cooker, sideboard, and sink with draining board. These working areas can be arranged in different combinations. Short distances to walk and fluent work sequences, combined with sufficient area to move around in, allow a high degree of functionality. Typical arrangements, depending on the area available, are kitchens with one or two runs of furniture, or U-shaped or L-shaped kitchens. > see Fig. 33

As well as this, free arrangements and central placing of cooker and worktops can be selected; these make the kitchen into a place with special qualities in terms of its function and style. > see Fig. 34 If a kitchen block is freestanding, it makes it easier to communicate with other areas or people, as users are not looking at a wall while working.

The kitchen is a functional place, a workplace within the home. Food should be easily accessible and often has to be stored in cool conditions. If utensils such as pots, pans and knives are well arranged and stored so that they can be seen clearly, this makes the individual working stages easier.
> see chapter Elements of Living, Storage

✎

\\Tip:
Ventilation is a key topic for kitchens, as cooking can produce very intrusive smells. For this reason, effective natural or mechanical ventilation should be provided.

◖

\\Hint:
Forward-looking kitchen concepts of recent decades, planning fundamentals and current trends are presented in detail in *The Kitchen* by Klaus Spechtenhauser (ed.), Birkhäuser Publishers, Basel 2006.

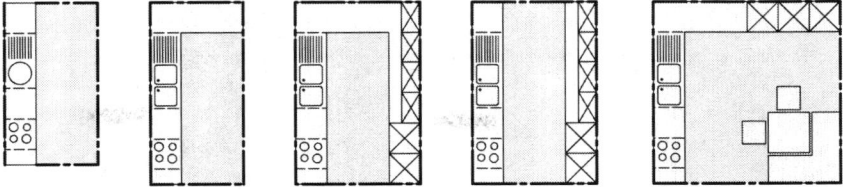

\\ Hint:
Generally, the depth of the movement area in front of kitchen furniture should not be less than 1.2 m, as otherwise people could get in each other's way. Worktops must be at least 90 cm high and adjustable above the plinth to match the users' height. The standard depth of kitchen cupboards under worktops, and fixed elements such as cooker and refrigerator, is 60 cm, thus giving the usual depth for a run of furniture.

\\ Tip:
Windows above the worktop permit visual contact with the outside world when working. But it is often difficult to adjust the sill height of such windows to the other windows when seen from the outside, as the height of the kitchen furniture means windows above worktops in a kitchen have to be correspondingly higher. If opening windows are placed behind a sink, the fitted height of the tops must also be considered, as otherwise it may be impossible to open the window.

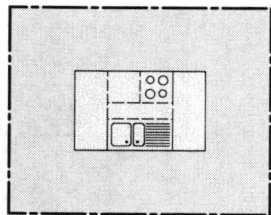

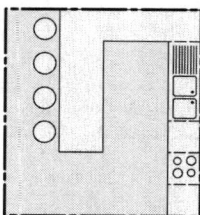

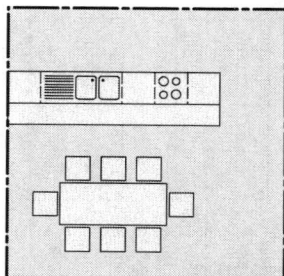

ANCILLARY FUNCTIONS

Uses and functions

Ancillary functions include washing, drying and ironing clothes. A special room is not needed for this, and other areas can be used temporarily to perform these tasks. But if it is possible to build a utility room for ancillary functions into the home programme, these activities can be combined sensibly. Like the kitchen and the bathroom, this room is a functional space earmarked for a particular purpose. For example, it can provide space for washing machine and dryer or washing lines. Here, laundry can be ironed and stored, along with household equipment and cleaning materials. A separate shower can do good service as well, so that it is possible to clean up when coming in from outdoors without impinging on more sensitive areas of the home. If the utility room is in a functional area together with the kitchen, they can be used at the same time and the distances travelled will be shorter. Depending on the overall concept, it can also make sense to place them near the bathrooms and bedrooms, where laundry is usually generated, and this avoids walking long distances to the washing machine.

STORAGE

Uses and
functions
People need a large number of objects in the course of a normal day. We surround ourselves with functional objects and personal memorabilia. Some of them are intended for daily use, and are well organized and stored functionally, while others have a special significance and are intended to be seen. A home is often a motley collection of a whole variety of objects and also a kind of personal museum that reflects the occupant's personality.

Storage
categories
We can distinguish between various storage categories. A private library imposes order on books while making them into exhibits; a collection of CDs or records makes it possible to play the desired music and not least expresses the owner's personal taste. These display forms of storage often have in common that they display personally important objects. A different approach is generally taken to everyday items such as clothing, shoes, kitchen utensils, cleaning equipment and comparable objects. They should usually be easily accessible, protected from dirt, and ideally invisible. They therefore tend to be arranged according to functional criteria and put away temporarily. Finally, a home also needs areas where things that are not in constant use can be stored efficiently over the long term. These will be objects that are seldom needed or are used seasonally, which would tend to be a nuisance elsewhere. In the field of tension between functionality and aesthetics, considerable significance is accorded to the places that display, organize and arrange things from various points of view.

> 📖

Room concepts
Beautiful and valuable objects, or those that are significant in some other way, affect and lend form to a room. Well-lit and proportioned areas of wall for pictures can be just as much of a design characteristic of a home as an open, room-dividing shelf system. An indentation in the floor covered with plate glass that can be walked on, or visual links with display niches and central gaps in walls, are all ways of presenting things in a home. It is also possible to create special rooms for collectors' items, or to use transit spaces as a gallery or library. > see chapter Elements of living, Circulation areas You do not have to be an art collector to know how to make

📖

\\ Hint:
As a rule of thumb, at least 2 percent of the overall space available in the home should be set aside for separate storage.

49

good use of this kind of design approach: it can appeal to many people who have things with which they like to surround themselves. In this context, quantity does not necessarily enhance the quality of a home, as living space can also acquire its characteristics from a few elements, which then make more of an impact.

Storing everyday objects Objects that are put away temporarily and used often require functional storage space. Small spaces for storing cleaning materials or food should be planned in relation to the areas in which these objects will often be used.

Cupboards, chests of drawers and shelves can be fitted in a kitchen or bedroom as needed, and in an appropriate style. › see chapter Elements of living, Sleeping, and Cooking

Long-term storage Storage space for the long term can be created in special spaces, in the cellar or attic if necessary. Spaces in the loft and under the stairs can also be allotted a function and activated, as can niches and corners. Good accessibility and efficiency are important. In the case of storage rooms that are not used very often, care should be taken to provide adequate ventilation where necessary, to avoid damp or stale air.

CIRCULATION AREAS

Uses and functions The indoor circulation areas in a home provide access to the areas dedicated to different uses. They separate and provide buffer zones between areas, but at the same time they connect them, and determine the routes taken through the home. They convey impressions of space, and their situation and volume greatly influence the quality of living. They need space, and just like all the other spaces in a home they need to relate correctly to building costs and designated uses.

Entrance areas in a home The hall is the entrance area to a home, and precedes all the other rooms. › see Fig. 35

In apartments, a hall is necessary for sound insulation purposes; it can also have some sort of climatic protection for the door. Natural lighting and the relationship between the area and height of rooms are significant here. › see chapter Basics, Creating space As the entrance area, a hall creates a transition from inside to outside, and provides a place where outdoor clothing can be left and guests received. According to the individual approach, it can be a minimal buffer zone or become an imposing entrance to the home. Without this important intermediate area, other use areas

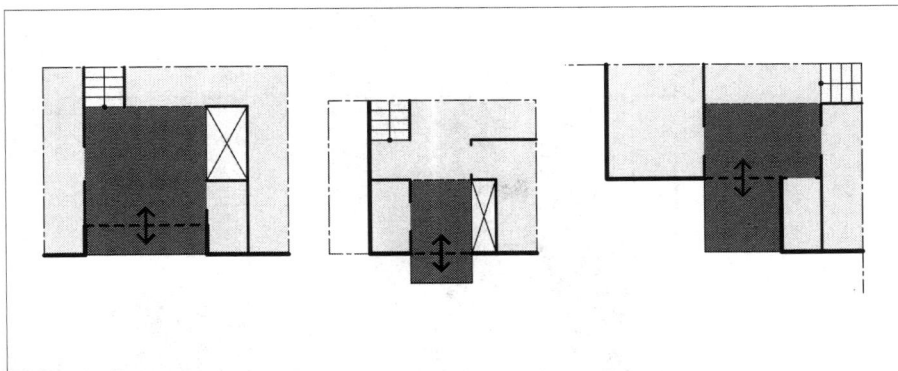

Fig.35:
Entrance areas to a home make the transition from inside to outside, meeting
criteria of function and space

may be disturbed or hampered. For example, the hall often provides direct access to the public living area, the kitchen, the cloakroom and the area where outdoor clothing is stored. In addition, it commonly also serves as a distribution area to other parts of the home and borders internal stairs and corridors.

Internal corridors

In hierarchically arranged homes, the corridors enable movement between areas used in different ways. If a corridor is not on the periphery of the home, and not naturally lit, it is an unattractive place in which to spend time: long, dark corridors have an unpleasant effect, and can generally be avoided by a clever overall design. Open ground plans make it broadly possible to manage without internal corridors, as the different areas merge into each other seamlessly. > see chapter Basics, Creating space

Transit spaces

If a corridor is naturally lit and covers a reasonably large area, its spatial quality is enhanced and it can serve as a temporary area in which to play and spend time, beyond its basic function. > see Fig. 36

Trapped spaces

Spaces accessed from a transitional area are called "trapped spaces" as they have no access structure of their own. Parts of a large room can also be portioned off with light, fitted structures (cupboards, screens or similar), so that the zone behind them can be used as a private circulation area. Seen in this way, an open ground plan consists of many transit areas, as the different zones run directly into each other and fuse with the access areas. > see chapter Basics, Creating space

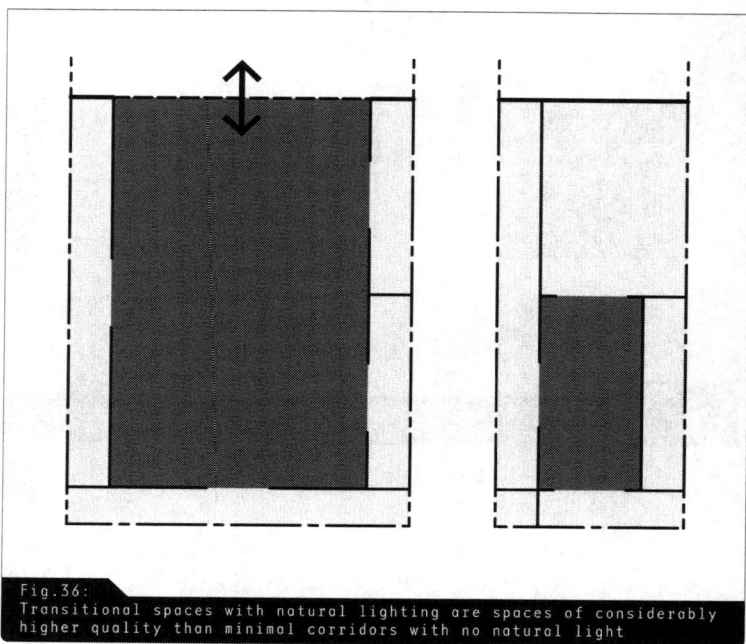

Fig.36:
Transitional spaces with natural lighting are spaces of considerably higher quality than minimal corridors with no natural light

Stairs

If a residential unit has more than one floor, stairs are needed. These stairs can be open plan, in another space such as the hall or living room, thus emphasizing the transition to the next level. Alternatively, it can be in a space of its own, screening off the upper floor. It must comply with functional and design principles, and those stipulated in building regulations. > see Fig. 37

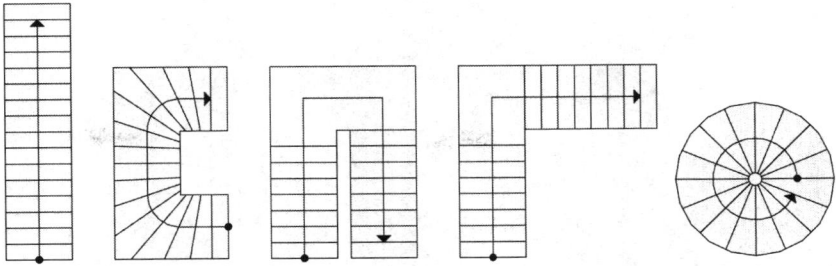

Fig.37:
Access to other levels inside the home using various staircase forms

BUILDING FORMS

The previous chapters dealt with the subject of homes mainly from the perspective of space and use. This chapter will categorize dwellings in terms of typologically different forms, the specific qualities of which affect living concepts considerably.

DETACHED AND SEMI-DETACHED HOUSES

Detached houses

Detached houses provide accommodation for a single family. They need to be a prescribed distance away from neighbouring buildings, according to local building regulations, and often require a considerable amount of space for garden and infrastructure. In comparison with an urban structure, these areas accumulate to reduce housing density. > see Fig. 38

Implications for occupants

Fundamentally, a detached house makes individual, and above all independent, living possible. Stylistic and other requirements can be interpreted on a personal scale in the design; there is a great deal of creative

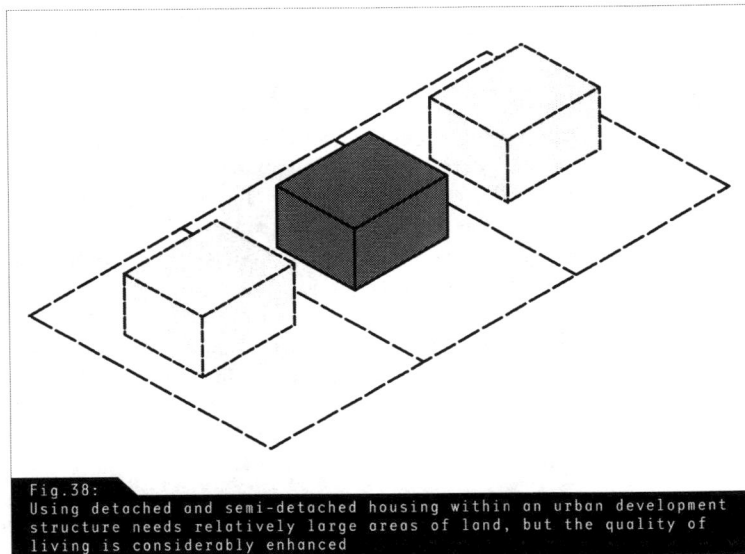

Fig.38:
Using detached and semi-detached housing within an urban development structure needs relatively large areas of land, but the quality of living is considerably enhanced

scope. These individual ideas, and the inclusion of a private outdoor space in the form of garden use, permit a wide range of design approaches.

Living on two levels

Single-family houses often involve living on two levels: the entrance floor commonly combines areas for general use like the kitchen, dining and living rooms. Areas for essentially private use such as hygiene and sleeping are then placed on the upper floor. But additional levels can also structure the space sensibly, according to need and the finance available. › see chapter Basics, Zoning, and Creating space

Bungalows

In bungalows, the use areas are arranged on a single level. Their low height means they may fit in very well in a country setting. The building is zoned within its own area, and can derive three-dimensional impact from projections and recesses, and create links with local features.

Semi-detached houses

Semi-detached houses combine two homes, each for one family, within a single building, and reduce construction costs by cutting down outside spaces. › see Fig. 39 The ground plans are often in mirror image from the middle, and vary only minimally, but it is possible in principle to build two different houses under the same roof. Changes of shape and material can be matched to each other, and convey a uniform impression despite different design approaches.

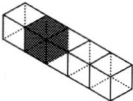

Terraced houses

Terraces are created by designing a row of uniform houses with identical ground plans or by accumulating individual built units. They are often conceived to save space, and are thus an economical way of building. They commonly occupy very little land, and restrict the width of the plot to the width of one house. › see Fig. 40

Such houses can be built in various ways: the fronts of the houses can be in a straight line, a circle or on the diagonal, and they can be arranged

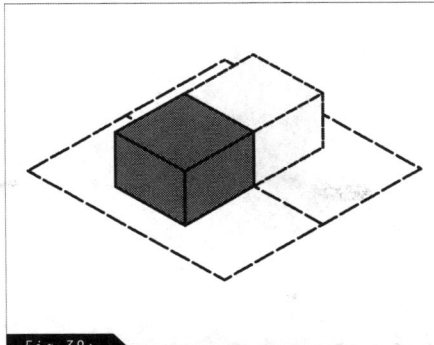

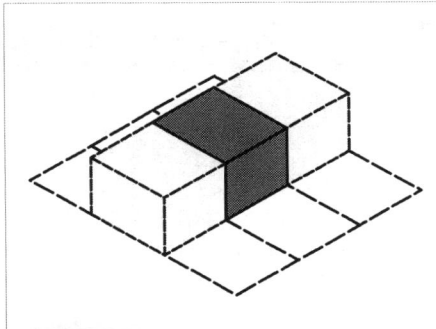

in rows or around the periphery of a block. Building terraced houses produces a high urban density that relates to a high quality of living and
the much-propagated development approach of a structured and yet not
unduly tight urban format.

Implications
for occupants
    The lighting possibilities are restricted in comparison with detached
and semi-detached houses: except in the first and last houses in a terrace,
the front and rear façades afford only two sources of natural light. For this

57

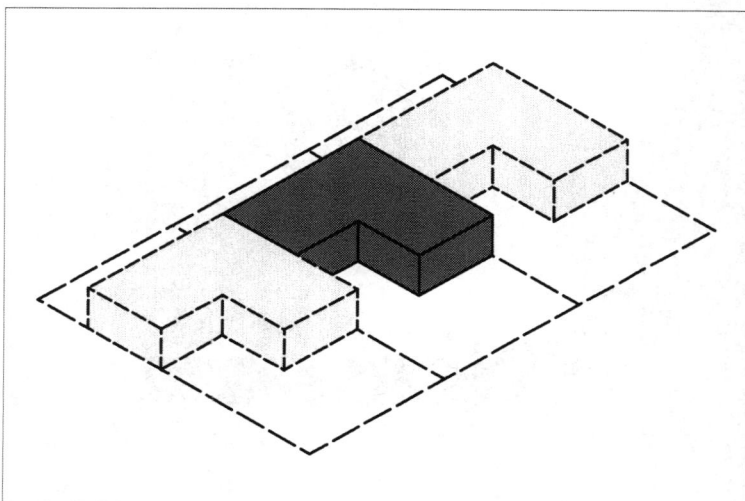

reason the orientation of the various rooms for particular purposes must be matched up very carefully. › see chapter Basics, Orientation Although terracing limits individual aspects of the outside space, independent approaches can emerge.

Garden courtyard houses

Garden courtyard houses are a particular form of terracing. Small private courtyards are formed using the end of the house next door, and these are largely protected from being overlooked. › see Fig. 41 This central outdoor space provides a meeting place for all the rooms facing it. As all the windows in the rear section look onto the courtyard, there may be a problem in creating areas for private retreat.

"Chain" terraces

To break up the visual impact of the façade, the houses are often staggered or made into chains by projecting or being recessed evenly. › see Fig. 42

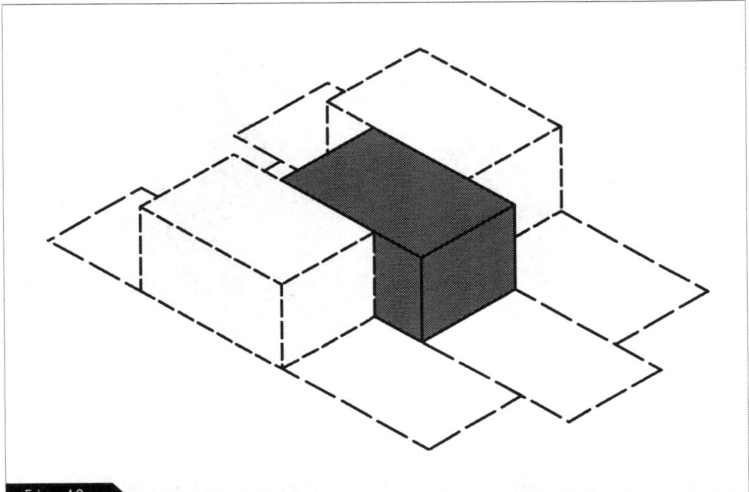

Fig.42:
"Chain" terraces create additional opportunities for natural lighting in the side walls as they are built with projections and recesses and form a sight screen from immediate neighbours

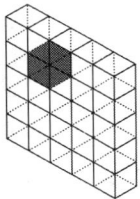

MULTI-STOREY HOUSING

Unlike single-family houses, multi-storey housing combines several residential units in a complex building, arranged adjacent to or above each other on several floors. Developing multiple levels on a relatively small amount of land increases urban density.

Block periphery development

Block periphery development is a closed construction form. A single building, or a series of individual buildings, encloses an inner courtyard and thus creates an interior area that is different from the outside. (see Fig. 43) A block periphery development can be constructed in a variety of shapes. Rectangular developments are just as acceptable as circular and curved structures, mixed forms or other geometrical figures. According to its size and shape, the enclosed inner area can be structured by other buildings and divided into additional courtyards. Inner courtyards permit all kinds of design approaches and opportunities for use. Possible access to the outside world through gates, passages and entrances offer public, semi-public and private use. Green spaces, designs such as urban squares,

> 🖋

garden zones and play areas are conceivable, and so are cafés, shops and
small public parks.

Implications for occupants

If the internal area is largely sheltered from the outside world, it
is also protected, in an urban setting in particular, from external influences such as noise, exhaust fumes and being overlooked. This inner
area is often semi-public in character, and is appropriately designed to
provide the occupants with green and play areas as a place for retreat.
Living rooms and bedrooms can face this quiet interior space, entrance
and ancillary rooms front the urban outside where possible, and form an
internal buffer against being overlooked and noise. But an urban block
can also simply frame an inner area, which then becomes a public sphere
by means of a raised ground floor, passageways and other openings, and
is able to invite urban life into it.

Building in rows

This structures a series of individual buildings or uniform blocks in
rows. > see Fig. 44 Although this could not be simpler as a basic form, the way
the various rows are placed in relation to each other can create a variety
of urban arrangements. Parallel, right-angled or diagonal arrangements
may be considered. Differentiated lengths and heights can create or adopt
spatial links within a complex structure. Building in rows does not separate inside and outside areas from each other as clearly as block periphery
development. If the rows are arranged entirely in parallel, the gaps created
are open at the ends. This affords good opportunities for natural lighting
and ventilation, but the development can be affected by noise, dust and
wind, especially in large complexes spread over a wide area. In addition,
the spaces in between have little inherent quality, as they are framed on
two sides only and are often subject to intense visual control over the
length of an entire row. One possible alternative is to close the open ends

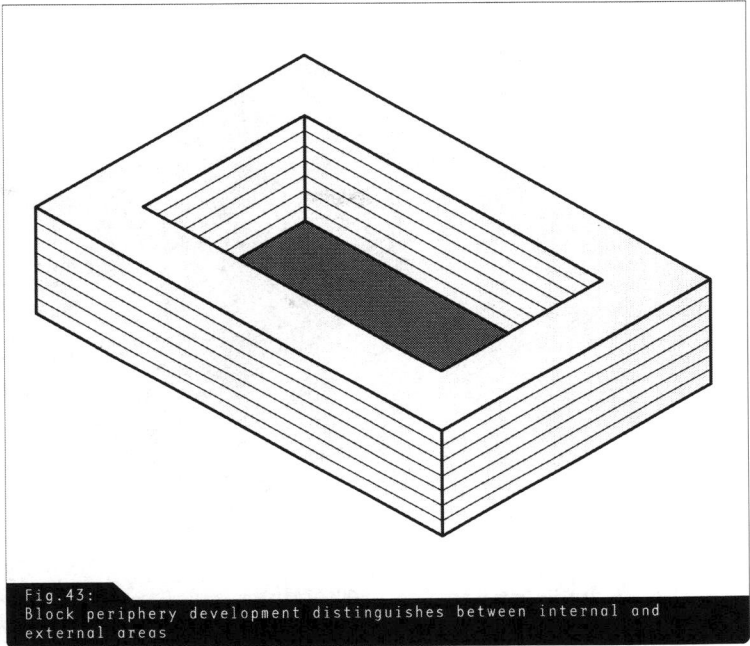

of a series of rows with more rows of buildings arranged at right angles. This creates internal areas that seem more closed in spatial terms and create semi-public, calmed areas.

Implications for occupants

Rows can be arranged in a clear east-west or north-south pattern, so that the areas devoted to various uses function well. › see chapter Basics, Orientation Unlike a closed block, no special solutions are needed for corners, which must be adapted to deal with particular aspects of access, natural light and ventilation. This means that standard ground plans can easily be implemented.

Block and row developments are usually characterized by coherent measures that put their stamp on a large area as urban development conceptions.

Solitaire buildings

In contrast, solitaire multi-storey buildings are usually planned with greater distances between the individual buildings, and it is quite common

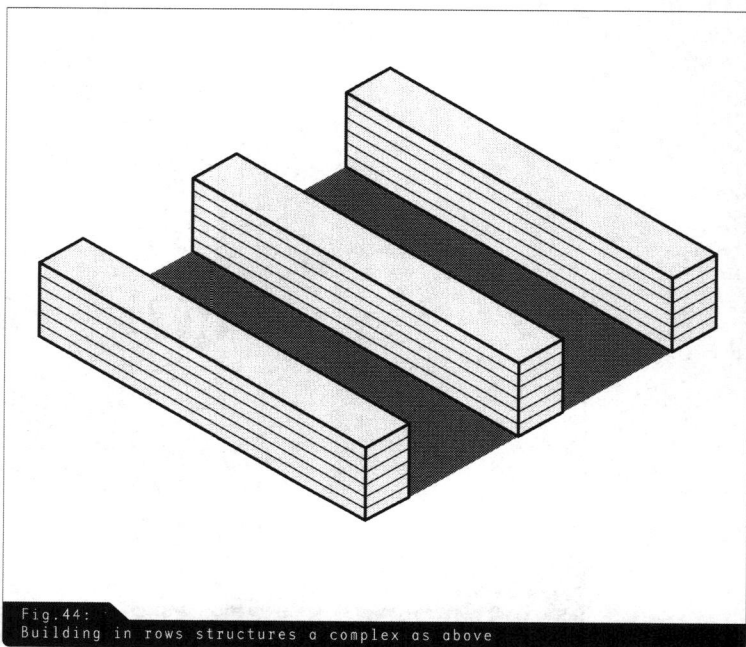

for them to be higher and longer. Such large dimensions can mean that a group of solitaire buildings do not produce coherent, spatially differentiated intermediate areas on a human scale.

Slab blocks   Slab blocks have a linear structure with a compact construction form that often reaches considerable heights and lengths. › see Fig. 45 left If combined into a communal complex, it is scarcely possible even for several slab blocks to create internal areas, so they often generate large intermediate areas that are not so simple to design.

Large forms   Slab blocks can be combined to create large forms, and as such can be planned individually or lend character to a large-scale development. › see Fig. 45 centre It is not impossible to differentiate space through a creative basic form and grouping, but it can be applied only to a limited extent because the areas involved are so large.

Point building development   Point building developments are made up of solitaire buildings placed freely within the surrounding area. › see Fig. 45 right Appropriate

ρ

\\ Example:
The 1947 Unité d'Habitation in Marseille,
designed by Le Corbusier, contains 337 two-
storey apartments in a building of 18 storeys
on an area of 138 m x 12 m. The ground floor
of the building is on pilotis, so that the
area it covers can be used as outside space.
The seventh and eighth floors provide space
for various shops and a hotel, and the roof
is planned to accommodate a kindergarten,
a sports hall and an open-air theatre.
Standardized serial production was intended
to combine efficiency with economic viability
and comfort for a large number of people.
"Stacking" living space and other functions
fitted in with Le Corbusier's vertical city
model.

design of their contours can impose three-dimensional form on the build-ings and thus underline their vertical quality. The undirected ground plan forms mean that the individual buildings have little relationship with the space around them, and because they are necessarily so far apart, it is impossible to create differentiated spaces in between even when they are placed in groups.

Implications
for the
occupants

> ρ

Solitaire building forms bring a large number of residential units together in one high-rise construction. Optimizing the ground and cir-culation areas makes it possible to create compact and complex housing structures that organize living space for large numbers of people within a relatively small area.

The living areas can be realized in a fundamentally individual way, or in a standardized form. But for economic reasons large-scale building schemes are usually realized uniformly, with standard ground plans that can be executed as various different types if need be, so that different dwelling concepts can be offered. General access structures restrict the de-velopment of individualized units, and emphasize the communal character of the building. The use of gardens linked directly to an individual home is possible only for ground floor apartments. Large or small balconies and roof terraces also create private outdoor space in multi-storey blocks. Ex-tensive views of the surrounding area can be an advantage on the higher

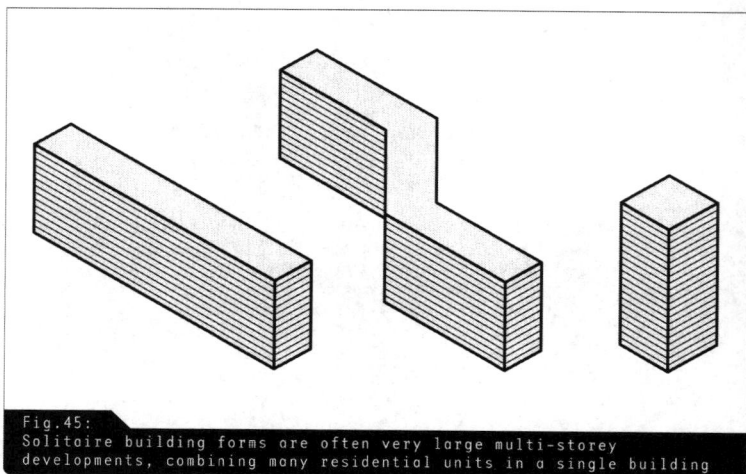

Fig.45:
Solitaire building forms are often very large multi-storey
developments, combining many residential units in a single building

floors. But the homes are exposed to particular weather conditions according to the level on which they are placed. High-rise dwellings can enjoy natural ventilation only to a limited extent because of their considerable height and strong winds.

\\ Hint:

Access is regulated by stairs, combinations of
stairs, and lifts and emergency staircases,
according to the height of the building
and the appropriate regulations. A lift
should be included for buildings of more
than four storeys, as the individual units
cannot otherwise be accessed suitably. High-
rise provisions within the local building
regulations stipulate the access and fire
protection requirements. These should be
addressed before drawing up a final design,
and they often have a far-reaching impact on
the way access and space is structured.

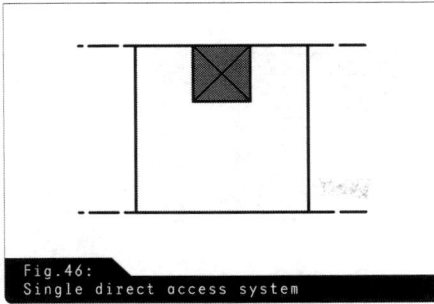

ACCESS

Solitaires and rows of detached and semi-detached houses all have their own internal access structures and independent entrances. This makes it possible to establish a sense of a personal address, and the individual use of these dwelling and construction forms is emphasized. Private access of this kind can also be implemented in principle in multi-storey homes, but generally means elaborate staircase structures are needed. Access systems for multi-storey apartment blocks are introduced systematically below. They bring general access areas such as building entrances, stairs and lifts together centrally and communally.

> 0

A typology can be established for various access forms. We distinguish between blocks with direct staircase access and blocks with corridor access. Both types are possible in principle for multi-storey blocks, according to the building form. However, corridor access makes economic sense only for projects involving considerable building length and uniform access systems.

Direct staircase access types

Blocks with direct staircase access allot a certain number of residential units to a central access area. The more homes per floor are accessed, the more economically the whole block can be built, as more occupants share an access point and there is a good relationship of circulation space to usable space.

Single direct access

Single access blocks provide access to one home per floor, and are essentially an uneconomical form, because a communal staircase provides access to only a limited number of homes and the costs fall to a small number of residential units. > see Fig. 46 It is usual to restrict such blocks to four storeys, so that no lift is needed. Advantages accrue from the freely

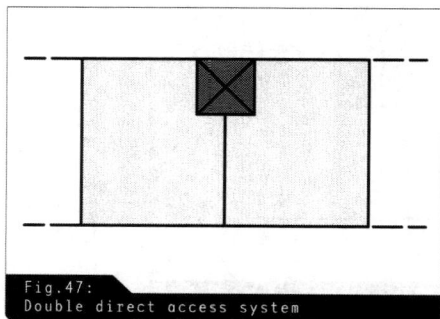

available lighting and ventilation possibilities, and in the design of the ground plans, as no constraints are imposed by other units on each floor.

Double direct access

Double direct access blocks provide access to two homes per floor and thus improve the ratio of access needs to use. > see Fig. 47 For rational construction the ground plans are often in mirror image on a central axis, but it is possible to realize different units in terms of room division and size. Advantages include the possibility of lateral ventilation from one side of the building to the other, and natural lighting on at least two sides, which permits the use areas to be favourably oriented.

Triple direct access

> 🗍

Triple direct access blocks provide access to three homes per floor. > see Fig. 48 Different dwelling sizes and plans take account of different user needs, and can thus enable a greater mixture of user profiles in a block. Depending on how the units are divided up, it can result in the ground plans facing one way, which creates constraints in orienting the different

🗍

\\ Hint:
Single and double direct access systems
can be readily supplied with fresh air by
lateral ventilation. This means a home can be
ventilated from one side of the building to
another, and thus all the air in a dwelling
is replaced in a short ventilation period.
A brief spell of intensive ventilation
guarantees a complete change of air and heat
loss is kept to a minimum.

66

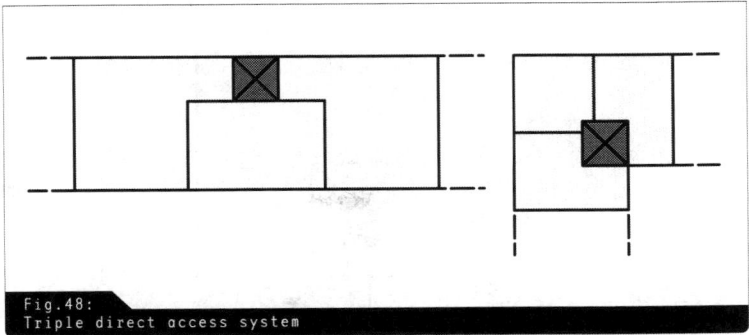

Fig.48:
Triple direct access system

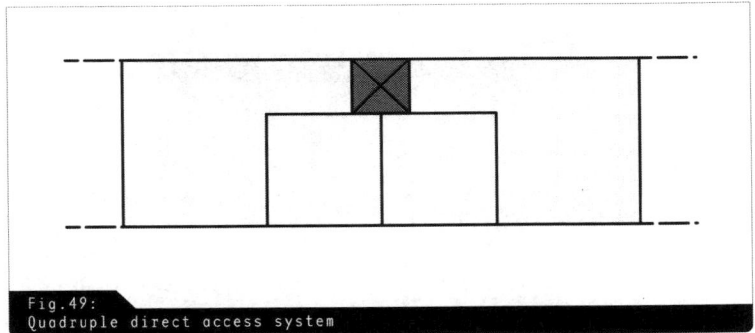

Fig.49:
Quadruple direct access system

use areas. The possibility of lateral ventilation is also limited. Triple direct access is particularly suitable for ground plans in the corners of the buildings.

Quadruple
direct access

Quadruple direct access allows simultaneous access to four dwellings on one level and is therefore an economical form. › see Fig. 49 The range of unit sizes and versions can be varied very considerably. Large and small units can be built on every floor. As in triple direct access, it can lead to single-sided orientation and restrictions on lateral ventilation.

Point blocks

As solitaire construction forms, point blocks cannot be arranged in series like direct staircase access types. They group the residential units by floor around a central, vertical access core. The number and size of the units are based on the floor area of the structure. In principle, units can face in two directions, which then offers opportunities for reasonable

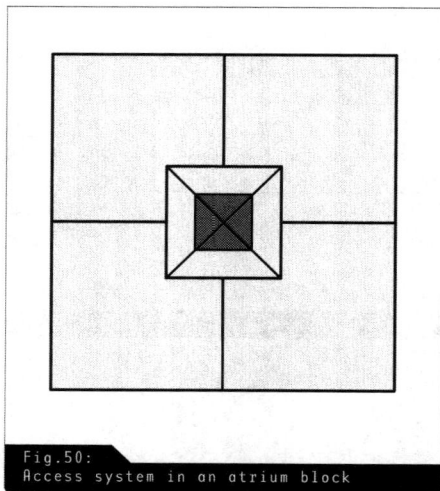

natural lighting and ventilation conditions – but only with up to four units per floor.

Atrium
buildings

Atrium buildings represent a particular form of point block in multi-storey construction. › see Fig. 50 Here, a light well admits natural light to the centre of the building, restricting its height, as sunlight can reach the lower floors to a limited extent only. An atrium makes it possible to avoid dark and unattractive access areas inside the building, and also creates semi-private zones in front of units, giving an impression of lavish space.

Corridor types

Residential blocks with corridors have vertical access that serves corridor systems, which in turn provide horizontal access to a certain number of adjacent residential units.

The vertical access areas can be placed centrally or structured sectionally, according to the length of the building. If the corridor is inside the building, it is known as an internal corridor, if it is outside, it is an external, or open corridor (covered walkway).

Internal
corridor

An internal corridor provides efficient access, as the outer skin of the building can then be used completely for lighting and ventilating the residential units. › see Fig. 51 Conversely, an internal corridor receives very

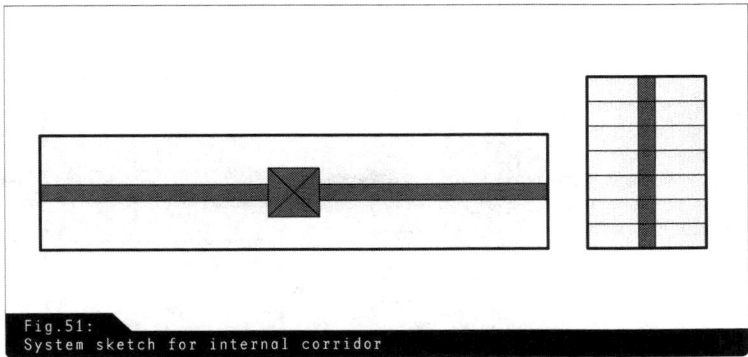

Fig.51:
System sketch for internal corridor

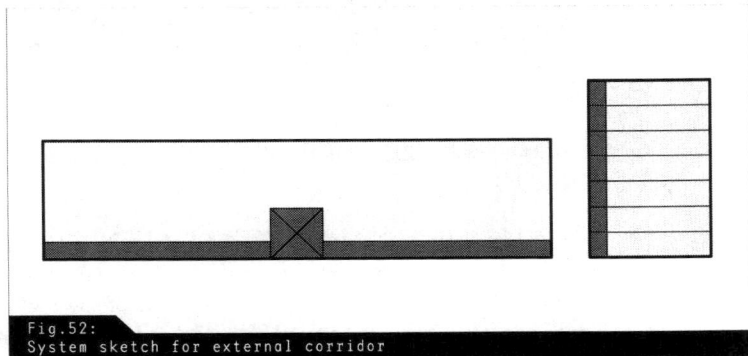

Fig.52:
System sketch for external corridor

little natural light. This means that such corridors are often dark and long, are uncomfortable to be in and tend to carry unpleasant associations. To avoid access situations of this kind, semi-public spaces can be created at reasonable intervals along the corridor. They break through to the outer façade and thus admit natural light to the middle of the floor. Such spaces accommodate vertical access cores like stairs and lifts, or are designed as waiting areas and meeting-points.

External corridor / covered walkway

Because they are placed on the outer wall of the building, external corridors have no lighting problems and can also be organized to be open on the outside, i.e. without glass or other separating structures. › see Fig. 52 However, there are climatic considerations, especially for tall buildings.

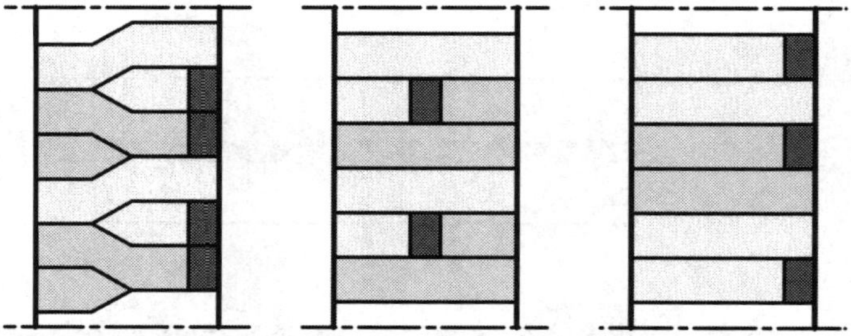

Fig.53:
Vertical staggering of access corridors can optimize the ratio of living to circulation space in maisonettes. The usable areas on the various levels can also be zoned and face outwards on two sides without access corridors

Wind and the risk of ice formation restrict the function of a covered walkway.

Vertical staggering

Essentially, every floor can have internal or external corridors. But an arrangement of this kind can mean ground plans are one-sided, as the corridors screen direct access to the outside of the building on one side, and even where they are external, only subordinate use areas can be placed on the corridor side. In a case like this, apartments and studio flats may be an attractive solution. If the homes are also connected vertically inside the building as maisonettes, access corridors are needed only every second or third floor, in internal or external corridor systems. › see Fig. 53, and chapter Basics, Creating space So for one thing the ratio of living to access space is optimized, and for another, lateral ventilation is possible through the full depth of the building, and the usable areas can face the outside on two sides on at least one floor. Of course maisonettes are not restricted to corridor access blocks alone. Direct stair access types and mixed forms can benefit from splitting a dwelling over several levels. Creating maisonettes is a good way of making two small residential units into one big one, particularly in old buildings, thus taking account of present-day requirements and needs.

IN CONCLUSION

"Building means designing life processes."

Walter Gropius

As an architectural discipline, housing construction offers almost unlimited creative possibilities. Designing key life processes can be associated with modern techniques and materials and a high level of comfort and style. The primal nature of the basic needs that are satisfied by creating living space can stimulate people to concentrate on essentials in a rapid and highly technological world. Housing construction intentions and methods have undergone considerable change in the course of history, but the most important parameter, the human being, has changed very little. It is still about creating appropriate, human living space that is capable of facing the future. New approaches and open thinking enjoy a similar status to looking back at the history of housing. Numerous ideas, insights and tried-and-tested concepts can be combined with fresh ideas, up-to-date forms of expression and new techniques. At present, housing quality is particularly associated with aspects of ecology and energy. In addition, demographic changes in our society are increasingly making an impact, and demanding appropriate housing. The benchmark for every potential solution in response to these challenges should be human beings and their basic needs, if the future is to be designed meaningfully. It is a matter not just of meeting the necessary housing requirements, but also of creating quality of life and accommodation, and continuing to question familiar and standardized housing typologies – above all, to create an environment that does not prescribe people's way of life, but in which they can develop creatively and design their individual ways of living together.

APPENDIX

LITERATURE

Andrea Deplazes (ed.): *Constructing Architecture*, Birkhäuser Publishers, Basel 2005

Klaus-Peter Gast: *Living Plans*, Birkhäuser Publishers, Basel 2005

Manuel Gausa, Jaime Salazar: *Housing / Single Family Housing*, Birkhäuser Publishers, Basel 2002

Dirk Hebel, Jörg Stollmann (eds.): *Bathroom Unplugged*, Birkhäuser Publishers, Basel 2005

Ernst Neufert, Peter Neufert: *Architects' Data*, 3rd edition, Blackwell Science, UK USA Australia 2004

Périphériques / IN-EX projects: *Your House Now*, Birkhäuser Publishers, Basel 2003

Friederike Schneider (ed.): *Floor Plan Manual*, Birkhäuser Publishers, Basel 2004

Camillo Sitte: *The Birth of Modern City Planning: With a translation of the 1889 Austrian edition of his City Planning According to Artistic Principles*, Dover Publications, USA 2006

Klaus Spechtenhauser (ed.): *The Kitchen*, Birkhäuser Publishers, Basel 2005

Marcus Vitruvius Pollio: *Vitruvius: The Ten Books on Architecture*, Cambridge University Press, Cambridge UK 2001

导言：家是什么？

"家"这个词涵盖了很多的需要和要求。家需要在同一场所内划分不同的区域，例如睡眠区、烹饪区、就餐区和卫生区。从这一点来说，家庭生活非常普通，但是你不能因此而觉得它不重要。在家里的生活是一个人确定自我的方式。一个人所选择的住所体现了他的喜好和条件。家的大小、形式和装修，以及它的使用模式会对居住其中的人的精神状态产生影响。这是一个避风港，但也是一个交流的场所；不管是对于性格内向还是外向的人来说都是如此。不同的事情可以同时发生，把不同的利害关系和功能互不侵犯地放在一起也是非常可能的事情。

家总是和个人的成长联系在一起。这一点在人类社会的个体发展过程中尤为明显：童年—少年—受教育—成家立业—退休—老年。通常这些生命阶段会有不同的需求，要求生活环境也随之改变。家应该尽可能地适应不同阶段的变化。这种可能性是非常有限的。往往搬家会比修饰、扩展或者改善现在的居住空间简单得多。有各种不同价格的、适用于不同年龄层次的房子出售，但是找房子往往是一件很麻烦的事情。我们对家所需要的东西有着固定的想法，只会因为经济或时间的原因而作出让步。

居住单元的设计不仅要满足必要的功能和相应的造价要求，还要创造高品质的居住条件。住户的需求是一个很重要的设计依据。如果这个设计是建立在个体的基础上的，如果未来的使用者是确定的，那么就可以直接考虑他或她的要求，而这个家就是量身定做的。如果没有特定的使用者，例如出租房或者开发项目，那么设计师必须按照一个虚拟目标群体的普遍需要进行设计。通常会为不同的使用者设计最灵活的标准组织平面，为个体的灵活性预留足够的空间，但是也要考虑住宅市场和长期的需求。

全世界已经有了而且仍然还会出现各种不同类型的住宅。地区开发要考虑不同的气候条件、当地的地形和材料之类的特点以及基本的文化状况。因此，沙漠地带和潮湿地区需要考虑的问题是完全不同的，城市和人口稀少的地区也有不同的要求，因此全球各地有许多种完全不同的住宅文化。这当然说明了很多住宅设计标准是有地域性的，但是有些基本的内容是可以应用在其他地方或者至少可以开阔设计师的眼界，释放出更多的潜能的。人是衡量过去、现在和未来的尺度。从这一点来说，回顾住宅的历史可以为现在的设计项目带来很多的想法和经过检验的概念。

当前的社会和住宅建设状况只是很短暂的一瞬，而未来的发展还很难预测。然而，居住建筑的使用寿命可以长达几十·年。因此，我们并不仅仅是为了今天的需求而建设的，而是为了子孙后代在建设。相对于美观来说，建筑的实际质量、功能和满足未来不同需要的可能性更具有决定性意义。某个特定社会形态的需要和要求是住宅是否能满足它的时代要求的基础。必须不断地检验这个基础能否创造出尽可能持久的住宅品质。

我们无法完全回答前面提出的问题："家是什么?"因为除了基本的要求之外，每个人对家和住宅都有他或她不同的理解。正是因为这个原因，感兴趣的读者可以问问自己家意味着什么，从而得出他们自己的概念，并且找到新的答案。接下来的章节就是为这些内容服务的。

基本内容

　　住宅设计在生活中创造了一系列受到大量内外部因素影响的事件。在这个语境下进行比较的一个重要标准就是人的基本要求、住宅的特定环境以及设计原则。下面总结了每个居住单元的基本要求。

生命周期

　　很多对家的要求都和人的生命周期有关，人的需要和要求会随着年龄的变化而改变。如果住宅的结构不能适应新的生活状态，它就会变得没有意义，住户就必须重新寻找一间适当的房子。因此，在最初的设计阶段考虑不同的生命状态有着重要的意义。理想的家应该可以适应儿童、老人和残疾人的要求，当然，这些要求是不可能同时满足的，而且也没有必要。例如，由于造价的原因很多多层住宅都没有电梯。但是可以采取很多实际的手段来进行弥补，或者至少可以有所准备：比如说，像开关、插座和门窗把手之类的服务设施可以安装在坐轮椅的人和孩子能够使用的高度。在设计的时候尽可能避免在同一层里出现不必要的高差或者狭窄的走道，因为美观的考虑可能会和后来的满足不同生活状态相矛盾。

　　无障碍是人性化住宅的一种特殊情况。无障碍建筑试图让每一个人都能使用某个场所或者某栋建筑，而不管他们的身体状况如何。不应该把身体有缺陷的人排除在外。从这一点来说，应该每个人都有无障碍的家，让所有的住户都能自在地生活。盲人或者行动不便的人、有其他缺陷的人、老年人和孩子尤其能从中受益。不是每一个家都必须设计成无障碍住宅，但是应该有改造的可能性。

　　从整体上来说，我们应该从寿命或者使用周期的角度来看待一栋建筑：大部分住宅会使用几十年，只有间断性的翻新或改造才能保证实际的质量，从而满足不断变化的需要。如果这些使用周期和因此而产生的时间段能够和使用者的生命周期相一致，那么就可以在设计阶段采取一些能够在之后的使用中很容易补充的措施。通过这种办法，不光可以从长期的经济因素方面，而且还可以从满足不断变化的要求和必要的改造方面来对使用者不断变化的要求加以规划。

朝向

　　自然光是住宅质量的重要标准之一。允许光线进入的或大或小的洞口让家有了很大的不同，而光是从东南西北哪个方向进入也是一个非常重要的因素。建筑的能量设计也会受到和太阳的相对关系的影响。可以根据不同的季节和气候区调节进入室内的太阳能，或者用遮阳设备来避免过热和眩光。后者可以用树木和植物、建筑密度、遮阳设施、阳台和挑出的屋顶来解决。但是在某些环境中，小窗户或者根本不开窗也许是最好的解决办法；同样的方法也可以应用在相反的气候条件中，在寒冷地区，很多热量都是从建筑北侧的窗户散失的。

地域特征　　诸如建筑、街道和开放空间之类的地方特征，以及特殊的地形和树木都会对居住建筑的设计造成影响，它们决定了总平面中各个不同功能区的朝向（见"基本内容，不同功能分区"一节）。因此，比如说交通干道的噪声或者相邻建筑的通视，都是可以在设计阶段通过用辅助房间遮挡比较敏感的区域或者适当地减少立面洞口来加以避免。相反，在有特殊的景观和安静的或者受保护的室外空间的地方，可以通过在立面上开洞把室外空间的设计和室内结合在一起。

使用时间　　住宅中不同功能分区的主要的使用时间和基本的采光要求是由房间和太阳路线的关系决定的（见图 1）。通过这个办法，建筑的朝向可以产生不同采光方式的空间，从而满足不同的使用要求（见"基本内容，不同功能分区"一节）。

罗经点　　在这种情况下，清晰的东西向或者南北向是建筑所希望的，但是如果认真设计房间的朝向，介于两者之间的办法也许能产生更好的结果。

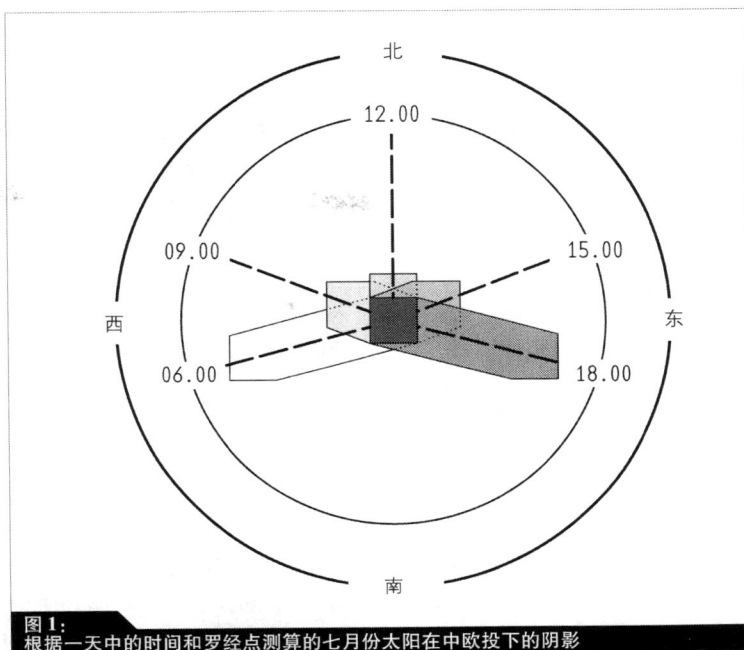

图1：
根据一天中的时间和罗经点测算的七月份太阳在中欧投下的阴影

北侧的特点是阳光和光线少。设计师可以把入口或者储藏室和辅助空间放在这里，因为它们通常不需要什么采光。太阳从东方升起，当它在天空中的高度较低时，可以给那些专门在早晨使用的空间带来阳光。这是厨房、成人卧室和浴室的一个很好的位置。南侧获得的阳光最多。儿童空间和起居室、露台和温室以及其他从早到晚都会使用的空间最好放在这里。太阳在西边落山，可以照亮那些主要在下午和傍晚使用的空间，比如说起居和休闲的空间。

P14

不同功能分区

住宅根据日常时间划分成不同的功能分区，它们构成了整个平面。

起居空间

起居空间是人们消磨时间的地方。它们的使用方式很大程度上和个人的需要和居住者的生活节奏有关。这些需要包括起居室和卧室，住宅中的类似空间可以单独使用（见"居住元素"一章）。

功能空间

诸如厨房、浴室以及可利用的专门的工作区之类的功能空间已经有了它们固定的用途。它们需要专门的给水和排水，只有改变建筑结构才能改变它们的用途。最早把房间变成住宅，让使用者能够免受外界影响的就是这些功能空间（见"居住元素"一章）。

交通空间　　内部交通空间，比如说走廊、门厅和楼梯，把住宅中不同的空间分隔和联系起来。交通空间的布置和设计会在很大程度上影响住宅的生活品质，因为它们决定了住宅中的房间序列，构成空间的层次。根据不同的设计方法，它们除了具有通道的功能之外还会有空间品质，可以暂时或者永久地发挥作用（见"居住元素，交通空间"一节）。

P15
分区

　　上面所提到的不同功能区——起居空间、功能空间和交通空间——在使用中是一个彼此独立又相互关联的矩阵关系，可以通过分区把不同的功能区组合成一个整体。因此，比如说卧室和浴室之类的私密空间（或许还带有一个前室），可以结合成一个整体（见图2）。厨房、餐厅和起居室的空间结构也可以组合成一个分区，这在住宅里更加常见（见图3）。

　　如果需要很多空间和不同的功能，那么可能在同一层解决就会比较困难，因为会受到通道和采光的限制，而且建筑的轮廓线往往不能随心所欲地布置。多层的设计可以把不同的功能空间布置在不同的标高上，从而分隔它们的不同用途（见图4）。例如，入口标高的空间多用于日常使用，而把私密的房间布置在不同的楼层上，通过中间的楼板避免空间之间的干扰。独栋住宅以及多层住宅中的公寓，可以做成两层或者两层以上的，分别布置不同功能的房间（见"基本内容、空间创造"和"建筑形式"等章节）。

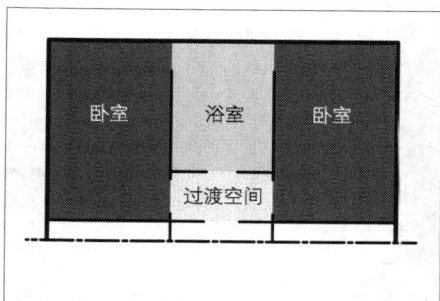

图2：
睡眠区的划分把私密空间组合在了一起

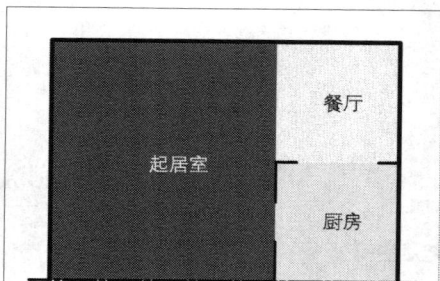

图3：
厨房、餐厅和起居室这样的常用空间可以布置在一个分区内

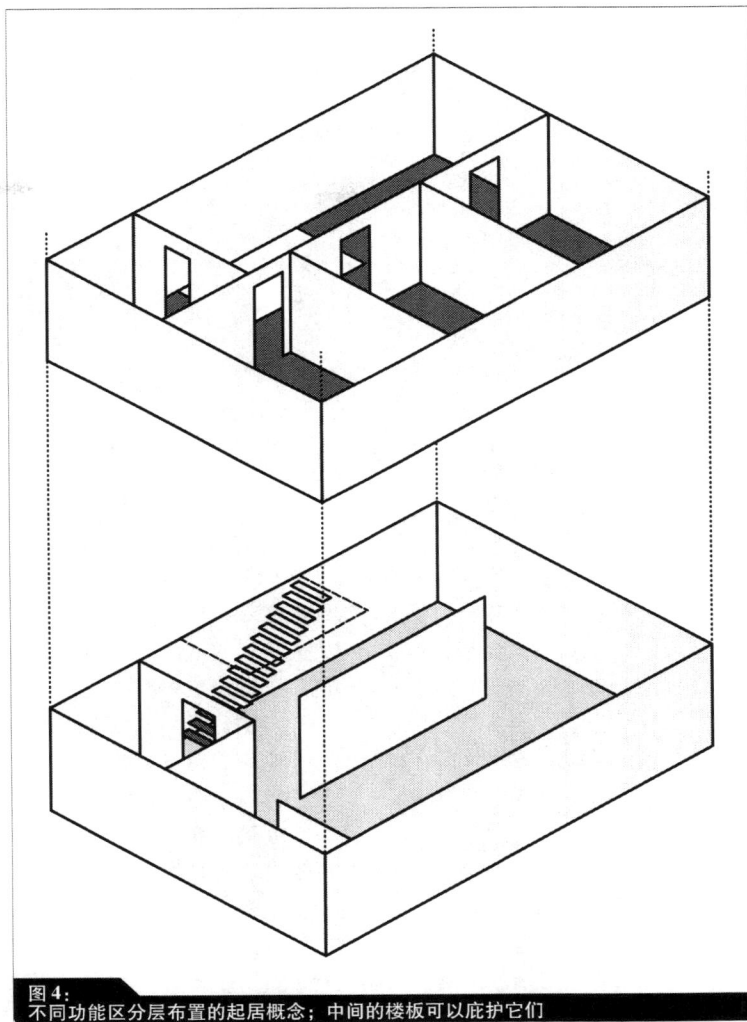

图 4：
不同功能区分层布置的起居概念；中间的楼板可以庇护它们

标准 　　住宅的分区标准可以是使用的时间段、特殊的工作区的需要，也可以是和使用者的需求相关的要求。不同类型的空间的分配有助于理清住宅中相关的品质或者是相对的使用要求之间的关系。因此，可以用诸如喧闹/安静，开放/封闭，以及内向/外向之类的特征来形成联系，使之彼此适应（见表1）。

住宅分区标准示例		表1
常用空间	↔	私密空间
工作区	↔	休闲区
起居空间	↔	功能空间
交通空间	↔	起居空间
成人	↔	儿童
与室外的联系	↔	与室内的联系
垂直交通	↔	水平交通
外向	↔	内向
开敞	↔	封闭
喧闹	↔	安静
明亮	↔	黑暗
白天	↔	黑夜

P17

空间创造

建筑设计创造空间和体量。用不同的功能分区创造出二维或者三维的联系，形成视线的通廊、空间的透视效果和交流的可能性。住宅中的房间受到总平面的形式和体量的限制。长、宽、高要满足各种需要和空间关系的要求。特殊的房间可以布置在一个特定的位置，并赋予其一个特殊的体量。与之相反，其他的中性用途的房间可以同等对待，这样就不会因为大小的不同而形成等级差别。另一个设计方法是把一个连续的空间或者一个完整的体量划分成不同的空间。这些不同的设计方法可以在整个设计中应用，也可以专用于不同的空间。

灵活性

从住宅建设来说，灵活性体现了单个房间或者整个居住单元的适应性。上面提到的中性用途房间的概念，以及后面将要提到的开放式平面，都是在使用和变化中保持灵活性的出发点。

为特定目的建造的房间

如果房间只是为了一个指定的用途而设的，那么它们就是为特定目的而建造的房间。在这里，空间、尺度以及它们和其他房间的联系是由专门的使用要求所决定的（见图5）。下面提到的要求是独栋住宅中常见的：起居室通常是最大的房间，然后是父母的卧室、孩子的卧室以及包括厨房、卫生间和辅助房间在内的功能空间。在这种情况下，

举例：

建筑师阿道夫·路斯的"Raumplan"（空间平面）概念严格按照功能、面积和体量处理不同的房间。设计体现了房间之间的独立性，但是从整体概念上看它们就像是一个连贯的空间结构；布拉格的穆勒住宅，1930年。

提示：

可以灵活划分、不受空间限制的灵活平面在结构上可以采用自由跨度的承重结构；在这种情况下，顶棚的荷载只分散到外墙或柱子，所有的内墙都是由可以灵活移动的非承重墙组成的。房间的分隔和边界可以由住户决定。

但是经验表明，这种住宅在实际使用中并不是大面积适用的：需要的时候，住户宁愿搬家。因此，这样的建造方法还会由于居住单元的大小而带来更高的费用。台地式的住宅经常采用经济的横墙结构，这样，墙与墙之间的空间就可以灵活使用（见"建筑形式"一章，台地住宅剖面）。

如果住户的要求变了的话，是很难改变房间的用途的，而且也不可能用很少的开支实现空间的重新划分。通常独栋住宅的整个概念是指向人的需求和愿望的，对于想要的住宅结构没有长期的改变。

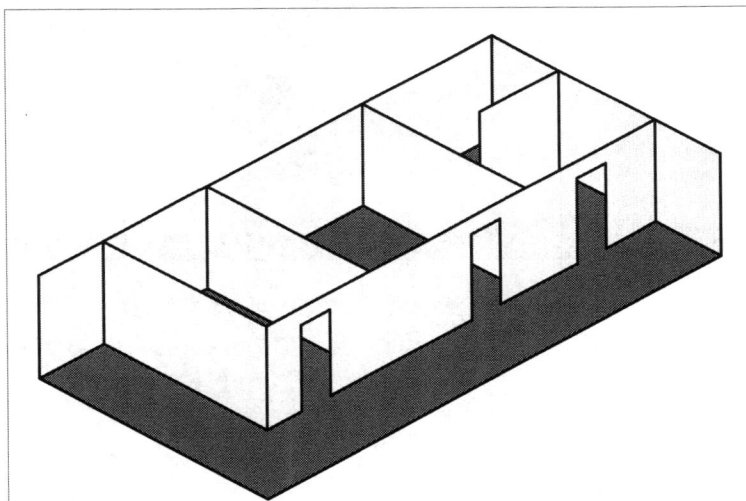

图5：
为特定目的建造的房间

中性用途的房间

如果房间的状况、大小和形状在很大程度上相似，它们就是中性用途的房间。这个办法对于出租房来说尤其适用，例如，那些在一段时间内会有不同住户使用的房子，因为我们不可能预知未来住户所有的需要或者变化。这些房间没有等级，可以根据功能互换。这个办法解释的问题之一就是为什么建于19世纪70年代的欧洲住宅经过翻修之后广受欢迎的原因，它们通常都有中性用途房间的设计（见图6）。因此，如果必要的话，可以让与功能相关的内部格局在很大程度上摆脱空间的限制。不仅是个人和家庭可以从中受益；现在不光是学生，还有越来越多的专业人员和老年人使用这样的住宅，因此都喜欢这种灵活性。

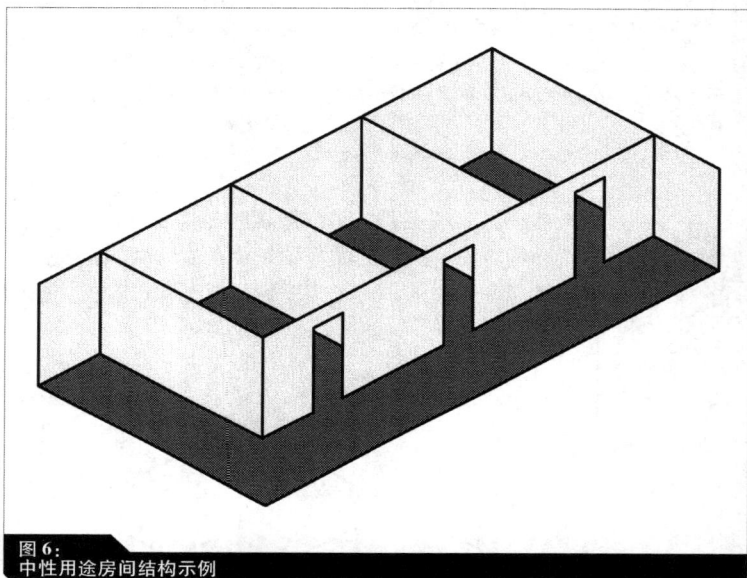

图 6：
中性用途房间结构示例

开放平面

开放平面指的是在住宅中通过流畅的过渡把各种不同的功能结合起来的平面形式。这个设计方法形成了宽敞连续的空间体量，没有复杂的通道和中间的隔墙。面积较小的住宅和公寓可以从中受益，因为空间不光用作通道。其他住宅形式所创造的界限分明的空间被开放平面中的分区所取代，这些分区可以通过改变材料或颜色、或者是用自然光和人工照明来加以强调（见"基本内容，分区"章节）。除此以外，开放平面还可以用架子、（透明的）推拉门和屏风等形式的临时或者（半）封闭的房间隔断来区分不同的功能，从而达到改变空间形象和视线的要求。

浴室和卫生间是惟一需要固定和隔离的功能区，但是特殊的设计方法还是可以模糊它们的边界。例如，它们可以结合成一个核，布置在开放平面的中间。这种布局可以在一个比较大的空间中形成结构上的空间划分，以满足不同功能分区的要求（见图7）。

除此之外，私密区和一般起居区可以跟厨房和餐厅结合在一起，使空间变得宽敞而易于管理。然而，在这种情况下，应该把入口区处理成室外到室内的过渡，这样就不会从室外直接进入起居区（见图8）。

82

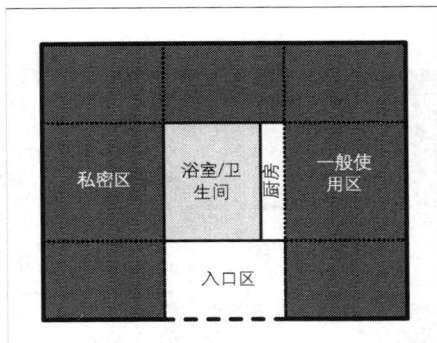

图7：
沿着开放平面中的功能核系统地布置使用空间

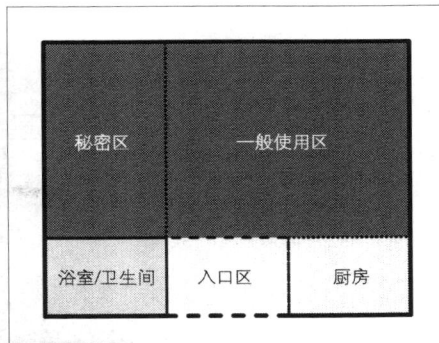

图8：
在开放平面中系统地综合和分隔的使用和功能区

图7中标注：私密区　浴室/卫生间　厨房　一般使用区　入口区

图8中标注：秘密区　一般使用区　浴室/卫生间　入口区　厨房

　　自由分隔使开放平面使用起来非常灵活，可以满足许多不同的需要和要求。近来很常见的单身贵族，以及年轻的夫妇家庭非常喜欢这种结构类型，开放平面只能在一定程度上适用于有不同要求的大家庭，因为噪声和功能交叉会造成相互干扰。

比例和房间高度

　　房间的高度是一种主观感受，它和房间的面积有关。房间越开放、面积越大，它的高度看上去就越低 (见图9)。

　　可以通过加大高度来强调某个功能分区，尤其是如果希望在一般的起居概念中把它强调出来的时候。因此单独强调某个重要区域是可能的，例如通过特殊的空间特征强调起居室或者餐厅，而把其他空间设计得相对平淡一些 (见图10)。

房间体量的不同标高

　　在一个体量中也能做出两层高的房间。例如，睡眠区可以塞进一个走廊，或者在一个公用的起居空间中创造一个比较高的后退空间 (见图11)。

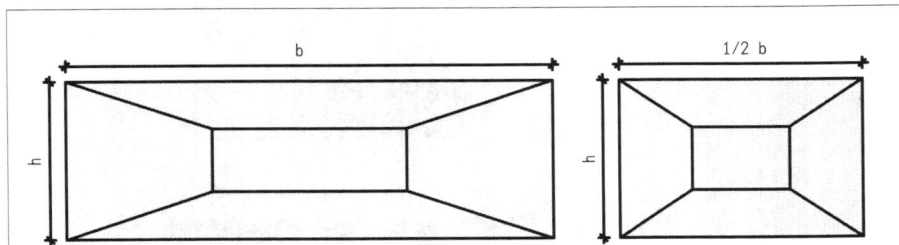

图9：
房间面积越大，高度看上去就越低

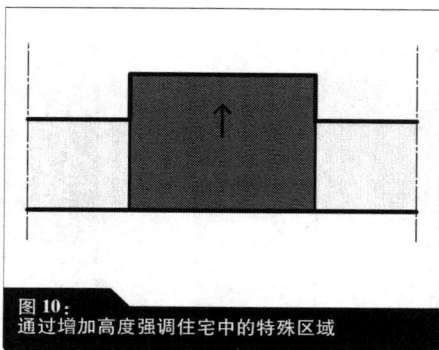

图 10：
通过增加高度强调住宅中的特殊区域

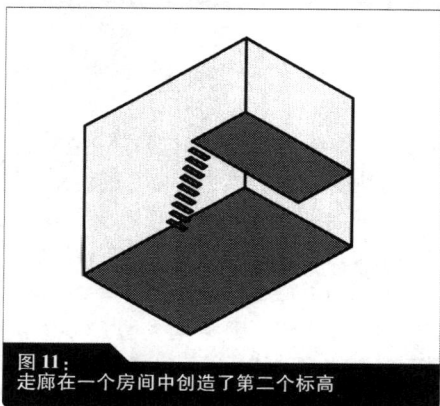

图 11：
走廊在一个房间中创造了第二个标高

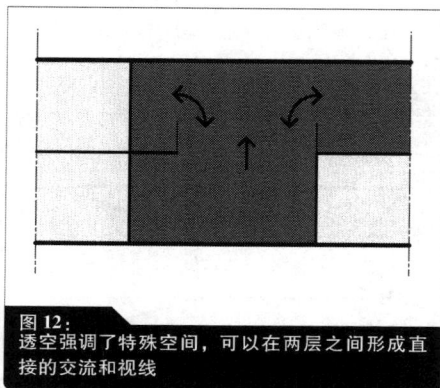

图 12：
透空强调了特殊空间，可以在两层之间形成直接的交流和视线

设计师并不是一直都有只从空间效果的角度选择楼层高度的自由，因为体量加大会带来额外的施工和采暖费用。例如对多层住宅建筑整体高度的规定会限制每个楼层的房间高度。

多层住宅

在一层以上的居住单元中，把两层或者三层结合在一起可以在空间和功能上创造出变化的空间效果。垂直通道有开敞楼梯和封闭楼梯之分，它们可以减弱或强化楼层的水平分隔。并不是只有楼梯能作为联系：在特定区域可以用透空表现空间体量。通常，通过附属楼梯与下面的楼层相连的走廊可以为楼层间的直接交流创造机会（见图12）。

复式住宅

在多层住宅结构中，所形成的形式被叫做小房屋——在一个住宅楼中的复式住宅（maisonette）（见图13）。这样的住宅特别能通过走廊

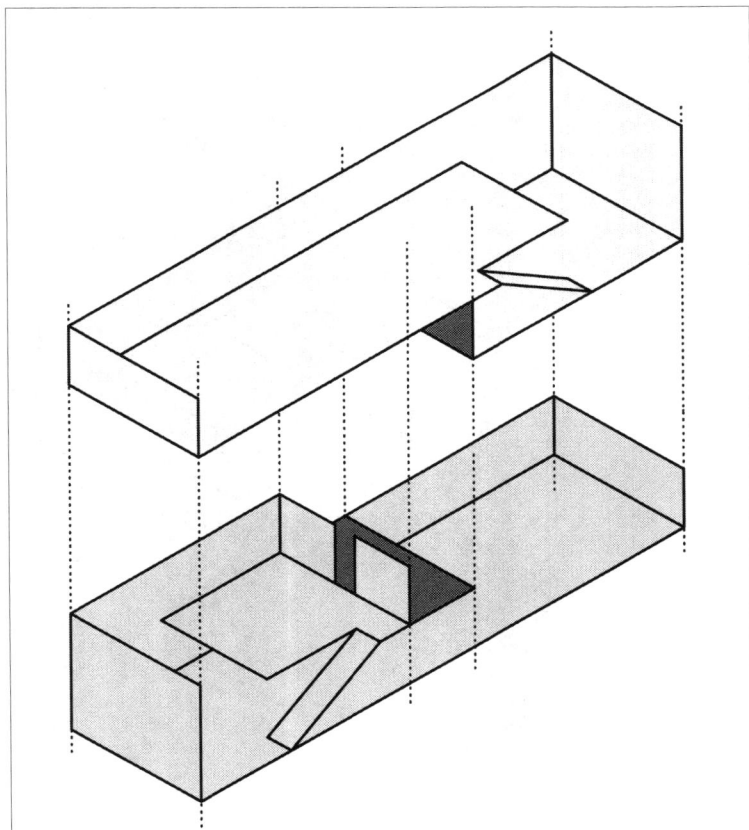

图 13：
小房屋可以在多层住宅的一个房子里形成两层或者更多楼层的效果

85

和透空形成特殊的空间效果。

错层住宅是复式住宅的一种特殊形式（见图14）。在这里，生活在只占据了建筑一半进深并用内部楼梯连接起来的不同标高上进行。这种布局方式意味着即使比较小的住宅也可以在不同的标高上进行划分（见"基本内容，分区"章节）。

从一个功能空间到另一个功能空间的过渡可以具有不同的设计特征，从而创造出不同的空间重点。没有固定边界的流动性过渡可以形成非常开敞的视野和连续的空间效果。反过来，墙体结构可以分隔不

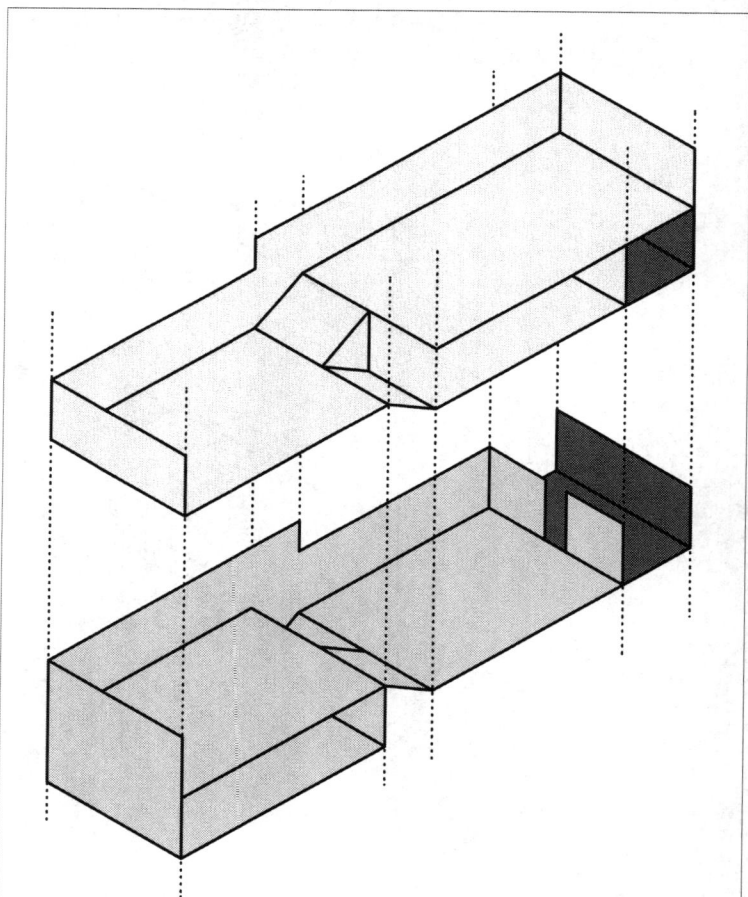

图14：
错层住宅在一个公寓内转换标高

同功能并明确地界定不同的功能分区。不同分区之间的联系洞口会影响空间变化的效果。洞口的宽度和高度有着同样的重要性。例如，通高的洞口会比相对较矮的门好。窄缝和宽阔的大门所形成的效果也是不同的。入口体现了功能空间的等级，表现了其内向或外向的性格（见图15）。推拉门和平开门可以灵活地处理分隔和联系。

不必给功能空间设计过多的洞口来形成通道。也可以仅仅是为了与其他房间的视觉联系而开洞，或者是功能性的开口。交通空间是无法单独发挥作用的，但是它可以强调空间的变化，确定空间的转换。

与室外的
联系

从整体上来说，室外空间是空间概念一个非常重要的因素。窗洞形成与地方特征相关的视觉联系，凶此在室内也留下了它们的印记（见图16）。视野会随着楼层高度的增高而变化。首层的公寓也许可以享受到便捷的交通和与花园的联系所来的好处，但是也带了室外通视的弊端。在必要的地方必须采取措施保证私密性。楼层越高，看得就越远，通视的干扰也越小。

出口、阳台和露台把室外和室内结合起来；室外空间可以成为起居空间的一部分。这样的联系可以通过墙体的延伸和棚架等结构形式，以及采用统一的材料来加以强调。

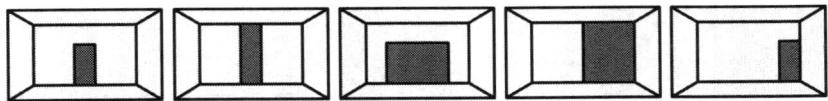

图15：
不同门洞对空间变化的影响，形成了入口类型的等级

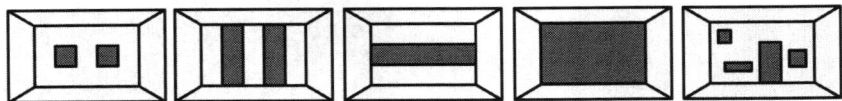

图16：
窗洞会在很大程度上影响与室外空间的联系，因此有利于赋予起居空间自己的特征

小贴士：
如果首层公寓外面的环境很嘈杂，可以通过几厘米的高差变化结合适当的窗台来减少被通视的危险。

居住元素

上面所提到的起居、功能和交通空间是住宅设计的关键元素（见"基本内容，不同功能分区"一节）。这些空间可以根据各自的需要和对不同住宅元素的常规要求来加以区别。

睡觉

睡眠是人的基本需要。环境的质量对睡眠会有很大的影响。这个特殊的环境必须考虑各种不同的要求，可以通过不同的方法加以区别。卧室主要负责休息和恢复体力，因此可以只考虑这些标准。这就意味着卧室是单一功能的、并且和其他功能分区完全分开的。

但是也可以在这个基本用途之上增加其他的功能。从而把它变成私密区和满足一天的不同时间中的私人要求的所在。

如果增加了其他用途，就要设计不同的连接通道（见图17）。如果设置了更衣间，那么最好让它和卧室直接相连，因为如果必须穿越中间走道的话会破坏它的私密性。步入式衣柜是一个明智的节约空间的选择。它最好与浴室相邻，避免穿越其他的公用区域。最好设计直接的通道，而不要借助于中间走道。

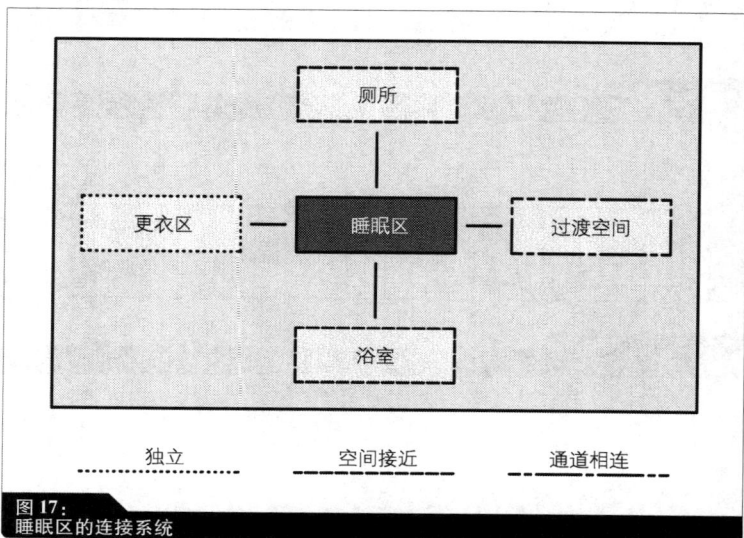

图 17：
睡眠区的连接系统

出于噪声的考虑，卧室最好与公用房间隔开。如果不能做到不同楼层上的水平分区，则应该增加过渡空间以避免与"喧闹"区的直接联系（见"基本内容，分区"一节）。这个建议在单身公寓中就不用采纳了，因为这里的使用只和单个个体的生活节奏相关，通常不会有其他人的干扰。

内向的睡眠区

睡眠区可以根据各人的需要用不同的方法进行设计。一个与内部和外部世界隔离、仅供睡眠使用的卧室是内向睡眠区的一种极端形式（见图18左）。容易对休息区造成影响的外界干扰被排除了。封闭的入口可以强调房间的屏蔽功能。尺寸和比例的变化可以形成宽敞的空间、或者甚至是如苦行僧般修道房间的印象，就像船舱一样——这是一个基本的例子。

与室外相连的睡眠区

把室外空间包括在内可以削弱这种内向的感觉。布置得当的窗户或者大面积的玻璃，包括朝向在内，都可以在很大程度上改变光线的条件，创造出特殊的空间效果（见"基本内容，朝向"一节）。从这个角度来说，窗洞是间接光线的来源，很难提供什么室外景观。特殊的窗户形状可以为特定视点的——比如说躺在床上——人提供一个外观世界的取景框。对这个想法继续深入，则可以在一个合适的、私密的情况下，把整个卧室的外立面做成玻璃的，这样可以大大地减弱室内和室外之间的界限（见图18，中间靠左和中间靠右）。

具有综合功能的睡眠区

根据通道和需要，卧室也可以设计得更加开放，以容纳其他的功能和用途。例如单独的工作或休闲区可以和睡眠区结合在一起，因为这些活动是在不同时间进行的（见图18，中间靠右）。

在外向的形式中，睡眠区可以布置在一个开放的楼层平面里，几乎和家里的其他功能没有什么区别（见图18右，以及"空间创造"一节）。

家具

和每个功能区一样，从根本上来说卧室也需要有足够的容纳家具、设备和交通空间的面积（见图19）。

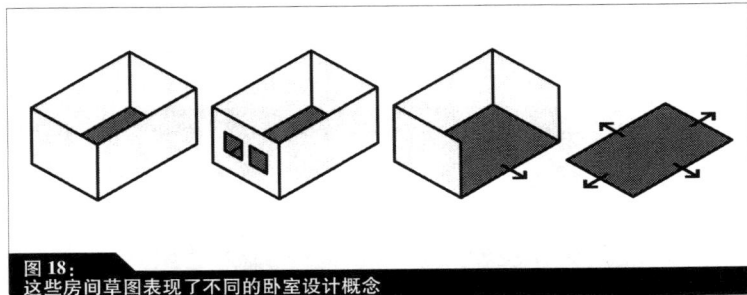

图18：
这些房间草图表现了不同的卧室设计概念

89

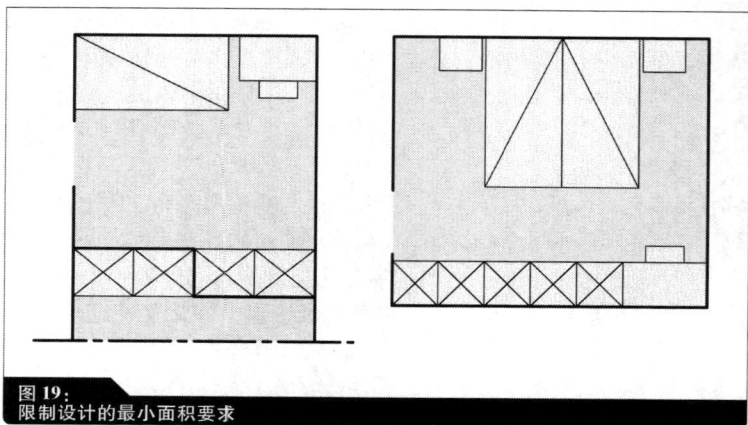

图 19：
限制设计的最小面积要求

这些空间的面积越小，设计的自由度就越小。优化空间当然是一种很好的方法，但是它只能在一定程度上发挥作用，还要允许在后续使用中有不同的设计方法。例如，标准尺寸的床确定了它需要多大的面积。但是中间尺寸的、自己制作的家具、有四根帐杆的床和抬高的床都不是按照标准尺寸制造的，而且都有可能成为某个睡眠区的一部分。

床的位置

床在房间中的位置会有意识或无意识地影响人的感觉。如果床布置在一角，就可以看见整个房间和它的洞口，体现了清晰度和安全性。如果床布置在中间，它就会成为房间洞口和其他家具的焦点，成为一个特殊的中心（见图20）。

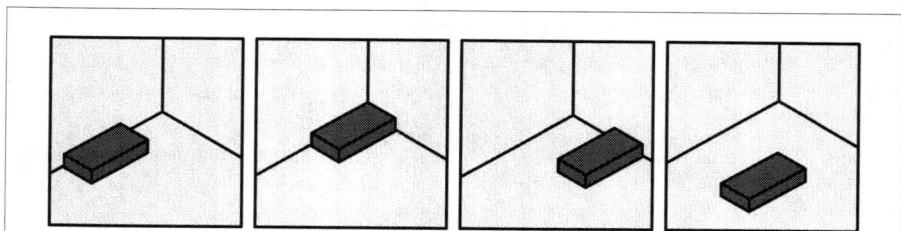

图 20：
床在房间中的位置会对安全感和躺在床上的人产生不同的影响

提示：

　　单个床的长度约等于使用者的身高加上25cm。放置床的房间要确保通达性。一般来说，一个人能够通过的宽度不能小于70cm。

通常睡眠区也是换衣服的地方，因此需要处理和存放衣物的空间（见"居住元素，储藏"一节）。如果没有设计单独的更衣室，就需要留有衣柜的空间。尤其要注意没有移动空间的房间，不应该把衣柜布置在门的轴线上，以免阻碍进入房间的通道。也可以采用嵌入式的通高柜子，因为它们是最好的利用空间的办法。但是这也束缚了墙体的面积和剖面，也会影响后来的其他设计方法。

吃饭

没有一种社交活动会像吃饭这样源远流长。它被上升到了政治和文化的高度，地方菜带来了地方和国家的自我意识，节日期间在家里或者在餐厅与我们的亲朋好友聚餐也是盛事一桩。但是更值得注意的是我们常常没有时间吃早饭、午饭或者晚饭，因为工作和休闲被认为更重要一些。一个家庭中的就餐区的状况取决于各人的要求。居住人口、就餐习惯以及可用的面积是非常重要的因素。就餐区不仅仅是基本的功能那么简单，它还是家人聚会和社交的场所。因此用紧凑、实用或者宽敞的就餐区来满足不同的需求。

通常厨房是为就餐区服务的，所以应该与之相邻。距离过长会给送餐和清洗餐具造成困难。如果就餐区在厨房的外面，可以用一个前面带有搁置物品的空间的窗口来改善操作过程。还要考虑接近前门和衣帽间，这样客人可以毫不费劲地找到桌子和厕所（见图21）。

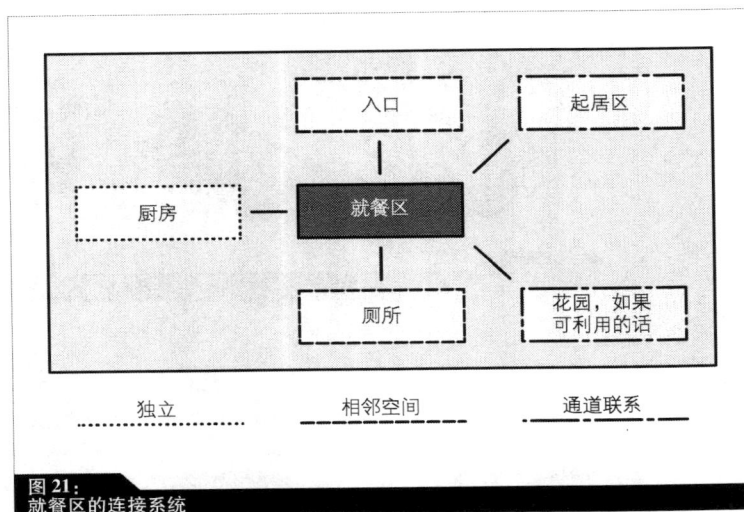

图21：
就餐区的连接系统

<table>
<tr><td>房间概念</td><td>可以把就餐区设置在一个明确界定的空间里，从而强调它在家里的重要性，以及它的独立性和独立的功能。如果这个房间还要供家人或者朋友的大型聚会之用，就应当规划适当的空间。另外也可以把就餐区布置在一个多功能的开放式起居室中（见"基本内容，空间创造"一节）。这种类型的空间没有明显的界限，通过把空间和功能结合在一起而形成开阔的视线联系，加强空间的宽敞感。另一个优点是可以灵活使用，因为必要的时候可以扩大桌子而不受空间的限制。如果想要在一个开放空间中强调出就餐区，最明显的办法就是加强空间内部的联系，用不同的材料和光线来达到效果。</td></tr>
</table>

房间概念

可以把就餐区设置在一个明确界定的空间里，从而强调它在家里的重要性，以及它的独立性和独立的功能。如果这个房间还要供家人或者朋友的大型聚会之用，就应当规划适当的空间。另外也可以把就餐区布置在一个多功能的开放式起居室中（见"基本内容，空间创造"一节）。这种类型的空间没有明显的界限，通过把空间和功能结合在一起而形成开阔的视线联系，加强空间的宽敞感。另一个优点是可以灵活使用，因为必要的时候可以扩大桌子而不受空间的限制。如果想要在一个开放空间中强调出就餐区，最明显的办法就是加强空间内部的联系，用不同的材料和光线来达到效果。

厨房中的就餐区

另一种可能性是把就餐区布置在作为起居室一部分的厨房里（见"居住元素，烹饪"一节）。这个办法减小了距离，保证了就餐区和食物准备区的交流。也可以在一个住宅里设两个就餐区，以区别厨房中功能性的快餐和早饭以及另一个用作正餐的空间（见图22）。

与外部空间的联系

住宅与外部空间的联系能够从根本上影响就餐区的品质。落地的窗户、出口或者（屋顶）平台可以在概念上把室外空间变成室内的一部分。光线以及就餐区的位置与太阳的关系也会对这个房间造成影响，能够对设计有所帮助，应该把一天中不同的时间考虑在设计中。（见"基本内容，朝向"一节）

家具

家具的布置和就餐区装修所需要的空间取决于空间的概念和使用的人数。

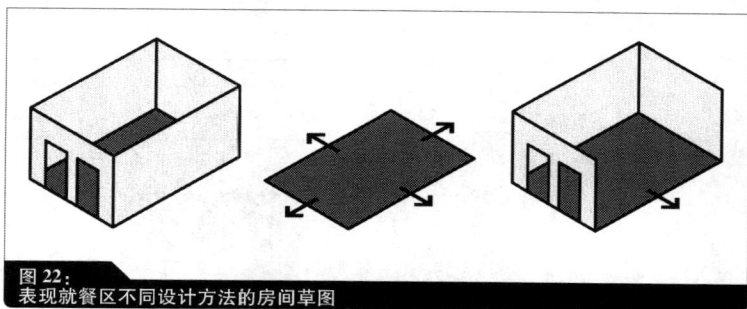

图22：
表现就餐区不同设计方法的房间草图

提示：
一个人在桌子上需要的最小空间为40cm进深，65cm宽，这样才能保证在吃饭的时候不会影响到旁边的人。供六个人使用的矩形餐桌的最小尺寸约为80～90cm宽，180～200cm长，圆桌的直径约为125cm（公式：以cm为单位的座位宽度×人数/314）。如果桌子在房间中是独立放置的，那么它和最近的墙之间的距离应大于1m，这样才能保证人坐下以后别人还能通行。

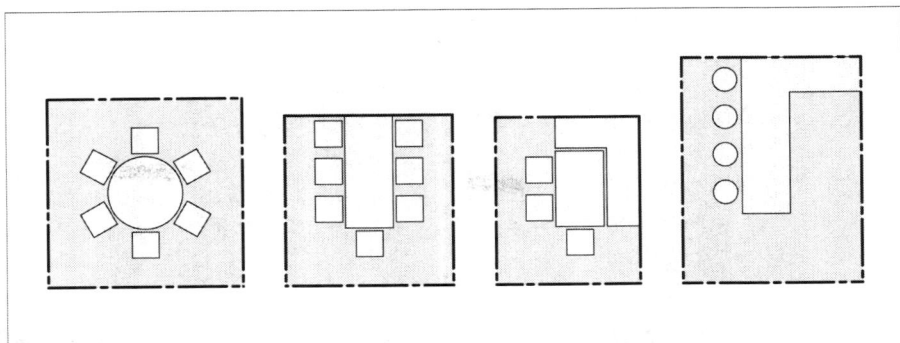

图 23：
强调功能或者空间的餐桌位置

　　餐桌可以有不同的布置方法，可以有不同的空间情况。圆桌能使整个场所变得均等，而矩形的餐桌会形成一种方向感。独立式的桌子会显得非常重要并且与空间结构建立起关键的联系。如果把桌子放在一角，有可能和一张转角的长椅结合在一起，那么就餐区就会显得非常实用。操作台也可以兼作厨房里的临时快餐台（见图 23）。

P35
用途和功能

　　工作
　　工作在我们的日常生活中占据了重要地位。它决定了我们的生活节奏、我们的环境以及我们的社会地位。我们的住宅条件也会——在本质上——受到工作活动的影响，因为它常常会决定我们能挣多少钱，但是它也会在功能上对住宅产生影响，因为不同的工作会对住宅提出特殊的要求。几乎每一个家庭都会提供一个脑力劳动的设备。尤其是对于通信和计算机技术来说，手机、笔记本和互联网让随时随地工作都变成了可能。因此卧室、餐桌或者起居室的沙发都变成了临时的工作场所。

　　但是由于各种各样的原因，我们很难在这些地方保证一个符合人体工学的长期的工作场所。特别是照明条件、噪声和功能的交叉使一个专用的家庭办公区显得非常必要，尤其是对于要经常工作的人来说。

空间概念
　　在住宅中设计一个工作区并且对它进行合理的分类和分隔，取决于住户的要求和所涉及的工作内容。例如，像簿记之类的常见办公活动，如果设计合理的话，通常只用一个很小的空间就能解决。比如说，如果这样的工作区布置在入口附近的话，就可以形成一个独立的接待工作来访者的场所，而不会对家里的私密区造成干扰。但是随着

93

要求的提高，工作区也可能会和开放平面中的其他功能撞到一起（见"基本内容，空间创造"一节）。很多职业都可以在家工作，根据不同的行为要求塑造不同的工作区。比如说，画家的工作室、雕塑家的车间、音响室或者音乐家的排练室，甚至是教师的书房都有特殊的、几乎不能与其他功能分开、而且允许工作和休闲之间互动的重点所在。在这里，分区是一种很好的解决办法（见"基本内容，分区"一节）。所有可能需要的与外部的联系，采光以及其他诸如隔声之类的标准都要根据特定的行为要求来决定。

娱乐休闲

很多休闲活动都是在一个封闭的空间中进行的，住宅中有很多娱乐休闲的空间。床不仅仅是供睡觉之用的，餐桌也可以用来玩游戏，浴缸可以是一个放松的场所，有电视和阅读区的起居室几乎被很多家庭看作是理所当然的一部分。几乎住宅中的每个房间都有利用作娱乐休闲空间的潜力。

起居区通常是住宅和就餐的核心区域，也经常被用作日常休闲的场所（见图24）。

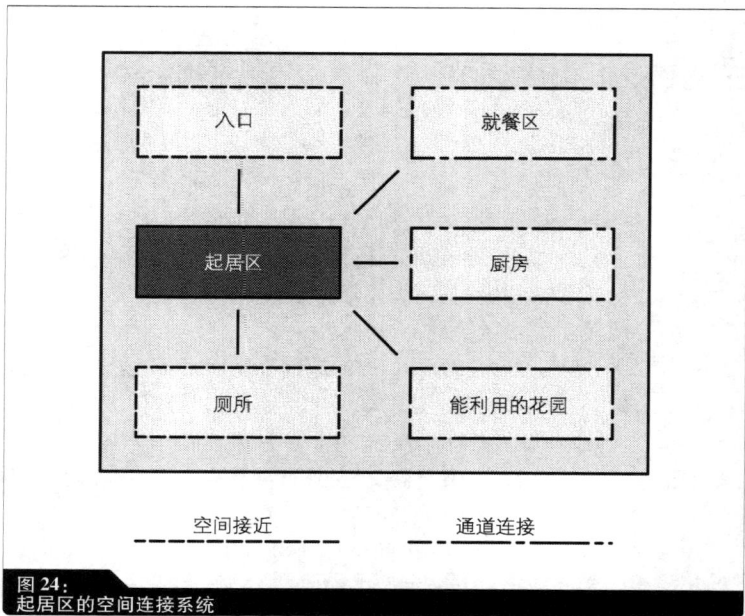

图24：
起居区的空间连接系统

94

起居室通常是住宅中比较外向的空间，是主要的消磨时间的地方，不光是主人，客人也能享受到它的品质。但是也有把内向的起居室做成住宅中一个私密的空间，与内部和外部隔离。

面积比较大的起居室给人以宽敞的感觉，可以进行二次划分，比如说分隔出阅读和游戏区、交谈和多媒体区（见图25）。

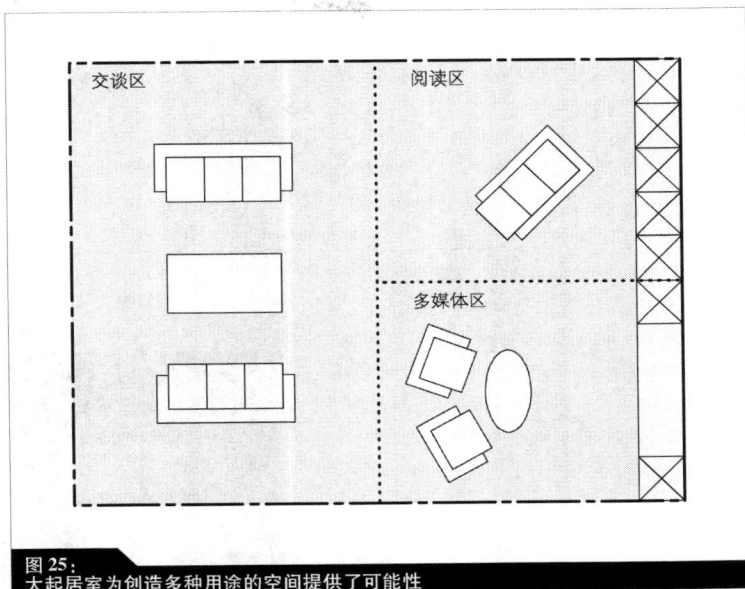

图25：
大起居室为创造多种用途的空间提供了可能性

娱乐区可以其具有特殊的空间特征，比如说，层高较高，或者在有些情况下通过透空和走廊与其他空间联系在一起（见"基本内容，空间创造"一节）。这些开阔的设计方法不是总能实现的，也可以用这种方法把比较小而紧凑的起居空间变成品质很好的休闲场所，尤其是强调生活中易于控制的、舒适的方面。还可以在（比较小的）起居室中增加其他主要功能空间，比如说就餐区和厨房，从而使开放空间的概念变得更加丰富（见"基本内容，分区"一节）。

健身区、音乐室、图书室或者其他单独的休闲区为创造性的设计和整体化的解决办法提供了广阔的空间。无论是临时的还是永久的使用功能，都可以按照住户的自己要求来确定。反过来，有的空间只能用于娱乐或者专门的休闲活动，比如说桑拿和私人游泳池。专用的儿童游戏区应该与其他功能空间有视线联系，以便大人在做其他事情的时候与孩子保持联系，并看着他们。从这一点来说，开放式的厨房餐厅、起居室和走道就比较理想：大人可以在附近的区域看到孩子（见"居住元素，交通空间"一节）。

把室外空间包括在内可以给住宅中的娱乐休闲区域带来很多的好处。例如，如果天气允许的话，首层的洞口可以把室外空间变成室内空间的组成部分，从而延伸起居空间（见图 26）。对于上面的楼层来说，大大小小的阳台和屋顶平台能够起到同样的作用。正如我们前面所说的那样，光线和与太阳的位置关系是非常重要的：建议用大的洞口迎接下午和黄昏的阳光，这样在起居室和休闲区的主要使用时间中都能够采用自然采光。

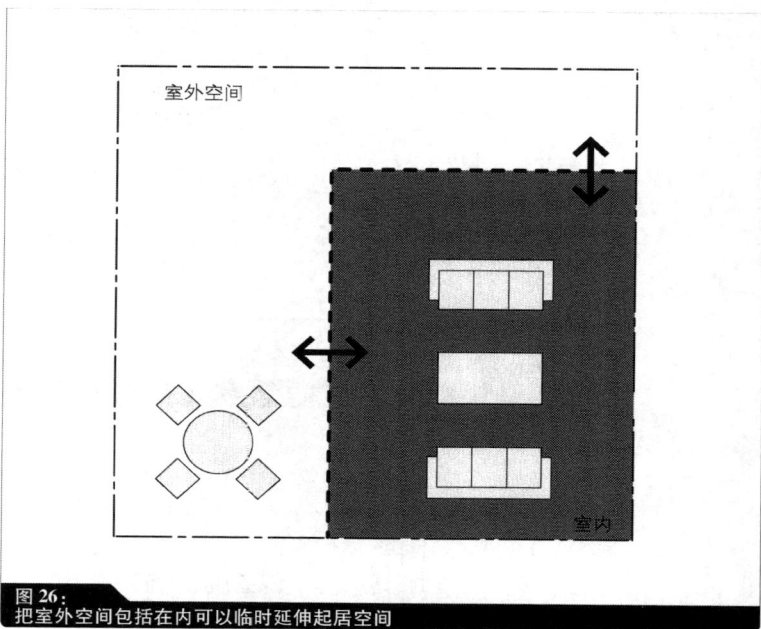

图 26：
把室外空间包括在内可以临时延伸起居空间

🖉

提示：

不用费多大力气就可以在住宅中给孩子们创造一个额外的空间。为了达到这个目的，我们可以试图通过孩子的眼睛去看住宅设计。小小的壁龛可以捉迷藏，靠墙而立的折叠桌在桌布的帮助下可以变成印第安人的帐篷，一扇小小的角窗可以在某个时间非常神奇地让太阳在瓷砖上闪闪发光。孩子是设计和即兴创造的高手。他们需要的一切东西都在他们的想像和房间中的小元素里。

卫生

生理卫生有很多表现形式。不管个人的清洁方式如何，它们对于生命和健康来说都是非常重要的。水是卫生的主题，而且在很多的文化传统中都占据了重要的地位。从整体上比较实际地来看，水必须以完全纯净的状态供给并且在使用后排放掉。必要的设施使浴室变成了住宅中非常有针对性的一个功能空间。和其他区域不同，它不能轻易地"转变"，必须体现不同使用者的要求。浴室和客用卫生间需要对视线和声音进行一定的遮挡，确保不对其他功能空间造成影响，或者被它们影响。

最好把浴室和私人卫生间布置在睡眠区的附近，以保证直接从睡眠区进入其中。如果可能的话，通向浴室的通道以及浴室区域，要与其他不那么私密的空间有视线的隔离。可能的话应该把衣帽间布置在住宅入口附近为公用的起居区服务（见图27）。

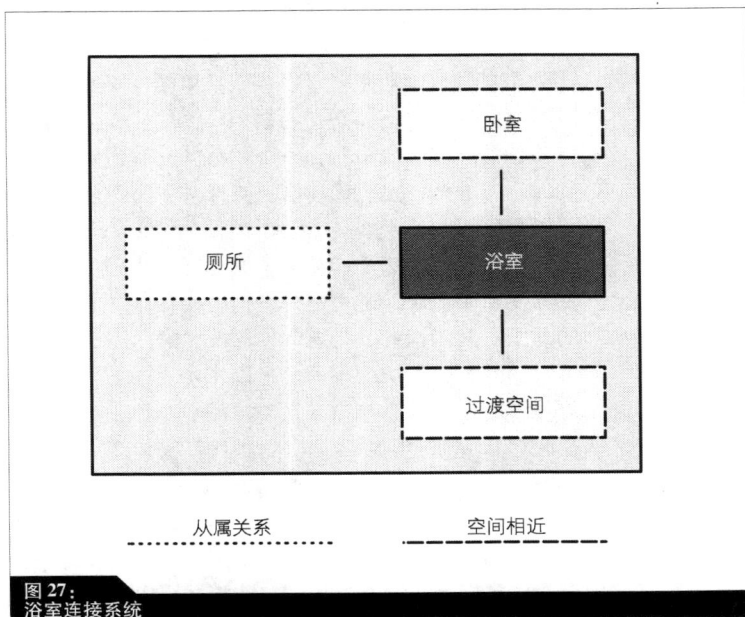

图27：
浴室连接系统

提示：

在设计的时候，尤其是设计一层以上的住宅的时候，固定上下水管的那面墙应该尽可能地凌驾于其他墙体之上，以减少安装的困难和噪声污染。在设计阶段就要考虑良好的隔声措施，尤其要考虑附近的住宅。例如，两户之间的管道墙可以彼此挨着，以免影响其他房间。为了使管线尽可能短，洁具应尽可能靠近管道墙。

如果住宅平面的大小有限，常常会把几个功能都组合在浴室中。厕所、浴缸、淋浴、洗手盆，或许还有妇洗器、给婴儿换尿布的桌子、洗衣机和烘干机都可以放在这里。浴室是非常普通的房间，面积小、功能性强。它的根本问题是那些组合在里面的东西是不是真的属于一起的。在上面的例子中，三个区域叠加在一起：个人卫生、洗衣服和上厕所。我们可以想想许多老的或者新的建筑中平面设计的卫生标准能不能达到功能和使用的要求。如果设计了一个辅助功能区，那么我们就可以把洗衣区放到其他地方（见"居住的元素，辅助功能"一节）。或者，也可以把洗衣机放到厨房、更衣区或者其他辅助房间里。如果能够把上厕所和私人卫生区隔开，浴室就可以变成一个愉快而整洁的放松和清洁的场所了。

房间概念

然而，就算严格运用这些标准也很难创造出高质量的空间（见图28）。至于是设计一个把淋浴、洗手盆和厕所都放在一起的节约型浴室，还是设计一个宽敞的浴室，这完全取决于可用空间的大小。

为了突出重点功能，可以在浴室里增加隔断之类的东西，把专用的空间隔离开，并且提高它的效率（见图29）。一个紧凑的私人卫生空间应该包括淋浴和洗手盆，一个单独的放置浴缸的区域。如果装有马桶和妇洗器的话，也应该和浴室隔开或者有专用的房间。这种布置方式可以避免相互妨碍和臭味，尤其是在早晨的时候。在比较大的住宅中，比较理性的办法是增加一个厕所（衣帽间）。

除了功能之外，浴室还可以被看作是一个非常令人愉快的消磨时间的场所。例如，可以脱离常见的洗手盆、浴缸和淋浴的模式，寻找一种特殊的平面和功能布局。可以用楼板标高上传统尺寸大小的水池来取代淋浴房。或者把（大）浴缸悬空放在屋子的中间，形成宽敞的空间感和额外的活动空间。这种设计需要有很大的空间，但是提高了浴室功能区的居住品质（见图30）。

与外部环境的联系

尽管浴室是一个非常私密的空间，但是也不能忽视它和外部环境的关系。如果这个外部环境是看不到的，那么设计一个大窗户或者小阳台会比较好。巧妙地选择洞口的形式可以在与室外空间建立视觉联系的同时，保证室内的私密性。

提示：

在德克·黑贝尔和约克·思陶尔曼编著的《无障碍浴室》（Bathroom Unplugged）中可以看到很　多富有创造性和前瞻性的浴室设计。Birkhäuser 出版社，巴塞尔，2005 年。

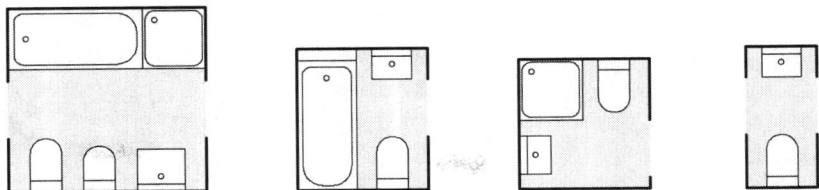

图 28：
节约型标准的浴室通常没有让人在其中消磨时间的吸引力

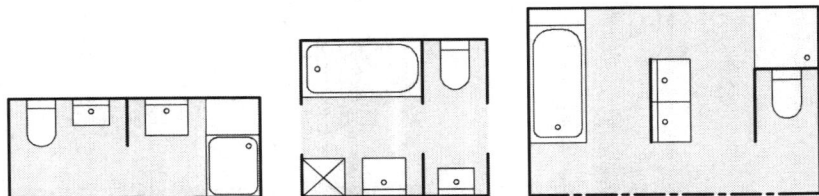

图 29：
根据功能划分的浴室可以供一个以上的人同时使用

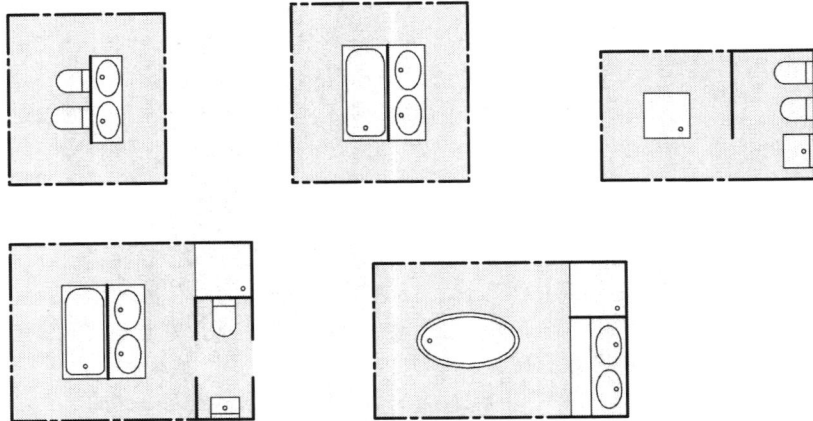

图 30：
浴室可以超越它们的功能进行自由设计

远离外墙布置的浴室无法与室外建立视觉联系或者自然通风，所以只能采取机械通风。单就通风问题来说，没有窗户的浴室不像有窗户的浴室那样能够吸引人在其中消磨时间。对于没有窗户的浴室来说，在屋顶上开天窗会是一个比较有意思的办法。它可以兼顾采光和通风等功能要求，而且还能从浴缸里清楚地看到天空（见图 31 右）。

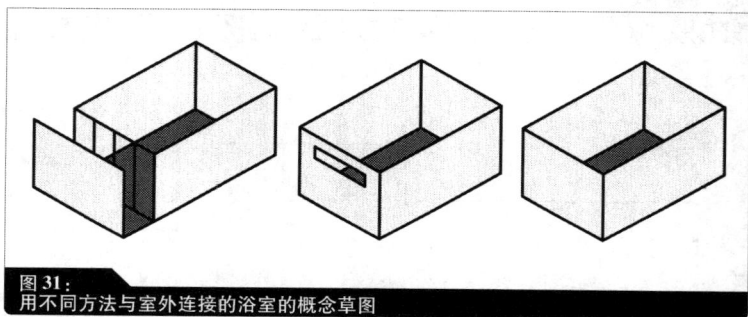

图 31：
用不同方法与室外连接的浴室的概念草图

烹饪

近年来烹饪已经失去了它的重要地位。人们不一定要为了吃而做。快餐和现成的食物决定了很多西方人的食谱，而批量化生产意味食品工业能够以最低的价格供应罐装的和冰冻的食物。但是烹饪并不仅仅是用尽可能短的时间准备食物这么简单。它还可以是一种具有很强休闲意味的公共活动，而不仅仅是加深我们对所吃的东西的认识。

厨房照理是住宅中最忙碌的一部分，因此非常重要。例如，它不能离住宅的入口太远，这样可以不用把买回来的食物提得太远。如果有储藏室和蔬菜园子的话，也应该离厨房近一些。厨房是为就餐区服务的，所以应该直接与之相连，这样可以毫不费力地把食物放到桌子上，把脏的盘子收走（见图 32）。

房间概念　　厨房可以有不同设计方法。它可以根据需要以不同的形式布置成烹饪式厨房、操作间或者餐厨一体的房间。如果把它当作一个独立的功能单元，就应该给它配备一个独立的房间，与其他的功能区完全分开。这个房间可以通过优化面积来发挥作用，也可以仅仅是一个操作区。但是它也可以是一个非常富有吸引力的消磨时间的场所，比如

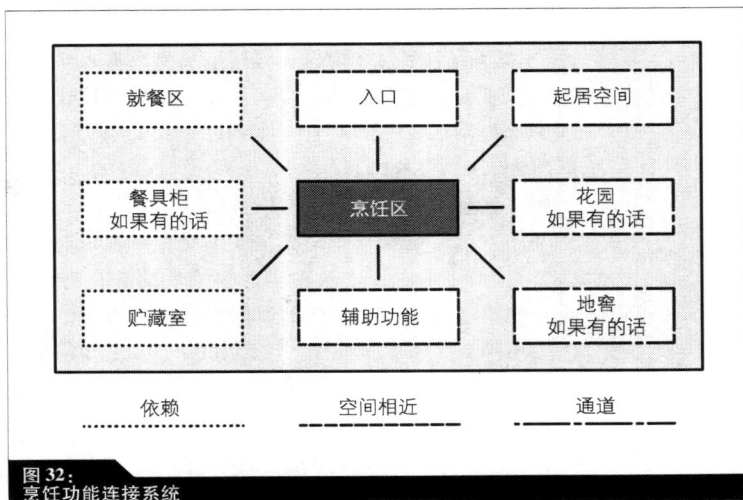

就餐区	入口	起居空间
餐具柜 如果有的话	烹饪区	花园 如果有的话
贮藏室	辅助功能	地窖 如果有的话

依赖 .... 空间相近 .... 通道

图 32：
烹饪功能连接系统

说，在厨房里设一个简单的或者完整的就餐区，使之成为家庭聚会的中心（见"居住元素，吃饭"一节）。也有人喜欢设计开敞式厨房，把它和其他的功能区融合在一起（见"基本内容，空间创造"一节）。通过这种方法，烹饪行为就能和其他的生活空间进行交流，并且能够从这个结合在一起的空间概念中受益良多。然而，值得注意的是，开敞的环境意味着做饭的味道有时候会对其他空间造成影响。为了灵活地解决功能和空间的问题，常常会用大扇的推拉门来作为临时的隔断。

与室外空间
的联系

对于厨房来说，与外部空间的联系是非常重要的，这不光是因为采光和通风的原因。比如说，对于有孩子的家庭来说，如果厨房的使用频率很高，能够看到在外面玩耍的孩子是非常有益的。在天气好的时候，附属的露台或者阳台是很好的早餐或者便餐的

小贴士：

通风是厨房的一个关键问题，因为做饭的时候会产生非常具有干扰性的气味，由于这个原因，应该提供有效的自然或者机械通风。

提示：

克劳斯·斯皮克腾豪斯编辑的《厨房》(The Kitchen) 详细地介绍了近年来比较具有前瞻性的厨房设计理念的设计要点和流行趋势。Birkhäuser 出版社，巴塞尔，2006 年。

场所。

厨房包括操作台、炊具、餐具柜和带有排水管的水槽。这些工作区可以以不同的方式进行组合。走动距离短、操作连续以及足够的面积能够提高厨房的效率。根据面积的不同，典型的厨柜布置包括单面、双面、U 形和 L 形几种（见图 33）。

除了这些方式之外，还可以选择自由式布局和把炊具和操作台放在中间的方法；这样可以使厨房成为具有特殊功能和风格特征的场所（见图 34）。如果采取自由式厨房，使用者能够很方便地与其他空间和人进行交流，而不是在工作的时候只能看着墙面。

厨房是一个功能房间，是家里的一个工作场所。食物应该很方便地取用，而且通常都要储藏在阴凉的地方（见"居住元素，储藏"一节）。

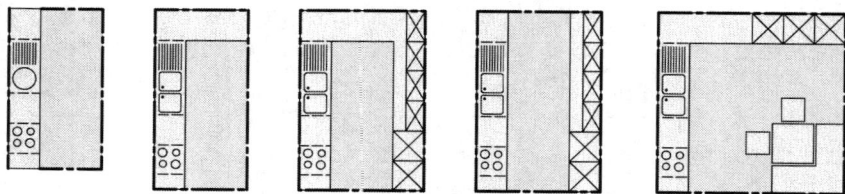

图 33：
简单的厨房和能够满足功能要求的最小面积的厨房形式

提示：

通常来说，橱柜前面的活动空间的宽度不宜小于 1.2m，否则就会挡其他人的路。操作台的高度应大于 90cm，并且可以根据使用者的高度进行调整。操作台下面的碗柜以及炊具和冰箱等固定部分的标准厚度为 60cm，这就确定了家具的常规尺寸。

小贴士：

操作台上面的窗户可以让人在工作的时候保持与外部的视线联系。但是通常很难把这些窗户的窗台和其他窗户的窗台在外立面上调成一致，因为橱柜的高度决定了厨房中操作台上的窗户要相应高一些。如果可开启的窗户布置在水槽的后面，必须考虑水龙头的安装高度，否则窗户就会无法打开。

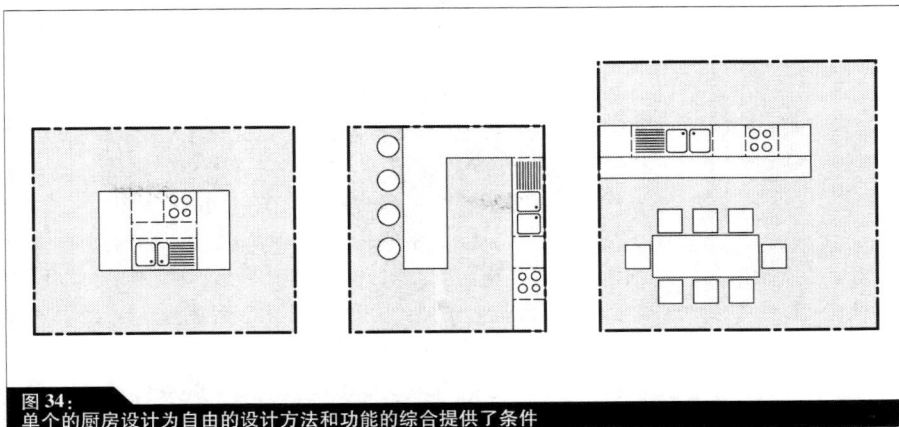

图34:
单个的厨房设计为自由的设计方法和功能的综合提供了条件

辅助功能

辅助功能包括洗衣、烘干和熨烫衣服。它不需要专门的房间，可以用其他空间临时替代。但是如果能够在住宅中设计一个功能性房间供这些辅助功能之用，就可以把这些活动结合得更好。跟厨房和浴室一样，这个房间也是具有特定目的的功能性空间。例如，它可以为洗衣机和烘干机或者是洗衣流水线提供空间。衣物可以在这里熨烫和储存，还可以放置其他家用设备和清洗材料。单独的淋浴也能起到很好的作用，这样从外面一进来就可以清洗干净而不用弄脏家里的其他空间。如果这个房间跟厨房在同一功能区，那么它们就能同时使用而且还可以缩短流线。从整体上来说，把它布置在浴室和卧室附近比较合理，因为通常那里有很多要洗的衣服，这样就不用走很长的距离去洗衣机那里。

储藏

日常生活中人们需要很多的物品。我们用各种实用的物品和个人的旧物把自己包围起来。有些是日常要用到的，放得井井有条，而有的东西有着特殊的重要性，是用来看的。家通常是各种物件的杂处之地，同时也是反映主人个性的私人博物馆。

储藏分类　　　　我们可以根据不同的类别对储藏室进行区分。私人图书室在展示书本的同时赋予它们一定的秩序；CD或者唱片架不光可以表现主人的品位，还有利于很快地找到想要的音乐。这些展示性的储物架一般都具有陈列重要私人物品的功能。另一个常用的办法是衣服、鞋子、

103

厨具、清洗设备和类似物品的收纳。它们要求取用方便、防止尘土、最好是看不到的。因此通常会根据功能要求进行放置，暂时地保存起来。最后，家里还需要放置不经常使用的、储存时间相对较长的东西的地方。这些东西或者很少用到或者具有一定的季节性，放在其他地方也许会成为不受人欢迎的东西。从实用和美观的角度来说，应该从不同的角度考虑那些展示、组织和放置物品的空间。

房间概念

美丽而贵重的物品，或者在其他方面具有重要有意义的东西，会影响一个房间的形式。挂在墙面上光线充足、比例均衡的地方的画可以成为和开放的、划分空间的架子一样重要的设计特征。地板上盖着可以上人的厚玻璃板的凹口，或者是展示柜和墙面上的中缝的视觉联系，都是在家里展示物品的方法。也可以为主人的收藏设计专门的房间，或者把过渡空间作为画廊或者图书室（见"居住元素，交通空间"一节）。并不一定只有艺术品收藏家才知道如何很好地利用这种设计方法：这对于很多有希望放在身边的喜欢的东西的人来说都是很有吸引力的。从这一点来说，数量并不一定能提升住宅的质量，因为生活空间也从一些令人印象深刻的东西中得到自己的特色。

日用品储藏

暂时存放和经常使用的东西需要实用的储存空间。存放干净的物品和食物的空间应当布置在经常使用这些东西的空间附近。

碗碟橱、抽屉柜和搁板可以根据需要以适当的方式布置在厨房和卧室中（见"居住元素，睡觉"和"烹饪"两节）。

长期存放

长期存放的储藏室应该有专用的空间，必要的话可以设置一个地窖或者阁楼。阁楼上面和楼梯下面的空间，以及壁龛和角落也可以加以利用。便捷和高效是非常重要的。对于不经常使用的储藏室来说，应该在必要的地方有足够的通风，以免产生潮气或者变味。

P50
用途和功能

交通空间

住宅中的室内交通空间是不同功能区之间的通道。它们在不同功能区之间隔出缓冲空间，又把它们连结在一起，同时还决定了住宅中的路线。它们传达空间的印象，它们的状态和体量会在很大程度上影响生活的品质。它们需要空间，而且和住宅中所有其他的空间一样，与住宅的造价和既定的用途有着直接的联系。

提示：

从经验上来说，至少住宅总面积的2%应该用作单独的储藏空间。

住宅的入口
区域

门厅是住宅中的入口区，先于其他所有空间（见图35）。

在公寓中，门厅是必要的隔声措施；它还可以在一起程度上起到阻挡室外天气影响的作用。在这里，自然采光以及房间的面积和高度之间的关系非常重要（见"基本内容，空间创造"一节）。作为一个入口区域，门厅是室内和室外之间的过渡，而且还提供了一个放置外衣和迎来送往的地方。根据不同方法，它可以是一个小小的缓冲区，也可以是住宅令人难忘的入口。如果没有这个重要的空间，其他的功能区可能会受到卡扰和妨碍。例如，门厅通常与公共起居空间、厨房、衣帽间和存放外衣的房间直接相连。除此之外，它还常常可以作为住宅中其他空间的分配空间，与内部楼梯和走廊相连。

内部走廊

在住宅的等级序列中，走廊可以实现人在不同使用空间之间的移动。如果走廊不是在住宅的周边，没有自然采光，那么它通常不会吸引人在其中停留：长长的、黑暗的走廊给人不愉快的感觉，通常可以通过巧妙的设计来避免。开放式平面为取消内走道提供了很大的可能性，因为不同的空间彼此间实现了无缝的连接（见"基本内容，空间创造"一节）。

过渡空间

如果走廊有自然采光而且面积比较大，那么它的空间品质就会得到提升，它就可以成为一个临时的娱乐和休闲的空间，超越它的基本功能（见图36）。

截留空间

需要从过渡空间进入的空间被称作"截留空间"（trapped spaces），因为它们没有自己的通道结构。也可以用灯光、固定的结构（碗碟柜、屏风或者类似的东西）把大空间中的一部分划分出来，这样它们后面的空间就可以作为私密的交通空间。从这个角度看，开放平面中包括了很多的过渡空间，因为不同的分区直接撞在一起，和通

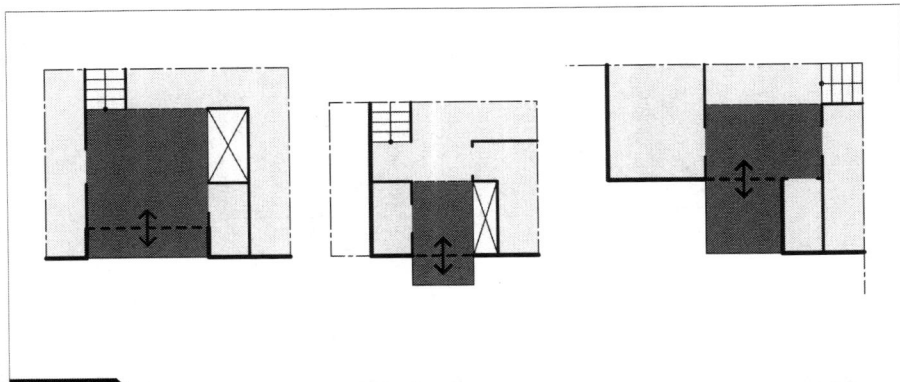

图35：
入口空间在室内和室外之间形成了一个过渡空间，可以满足功能和空间的要求

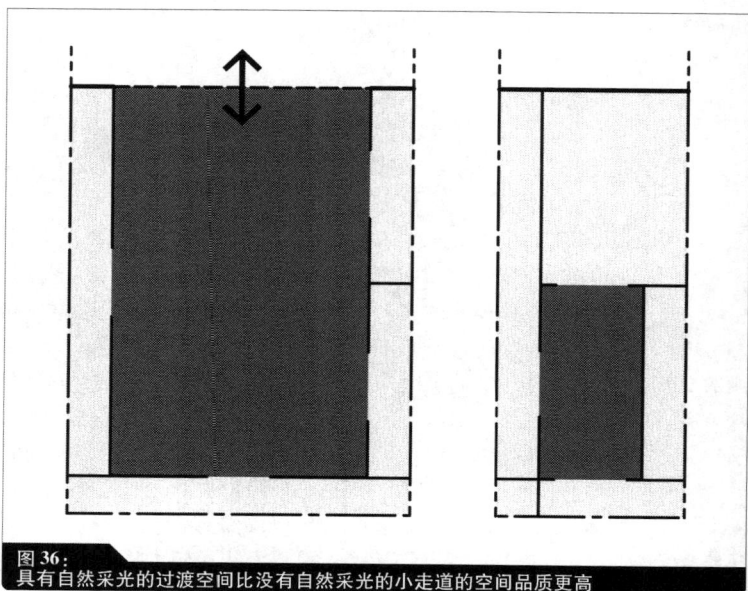

图36：
具有自然采光的过渡空间比没有自然采光的小走道的空间品质更高

道空间融合在一起（见"基本内容，空间创造"一节）。

楼梯

如果一个居住单元的楼层超过一层，就需要有楼梯。这些楼梯可以是开放式的、位于门厅或者起居室之类的空间之内，从而强调和另一个楼层之间的过渡作用。也可以是封闭式的，与上面的楼层隔开。楼梯的设计必须把功能和设计原则结合在一起，并且遵守建筑的其他规定（见图37）。

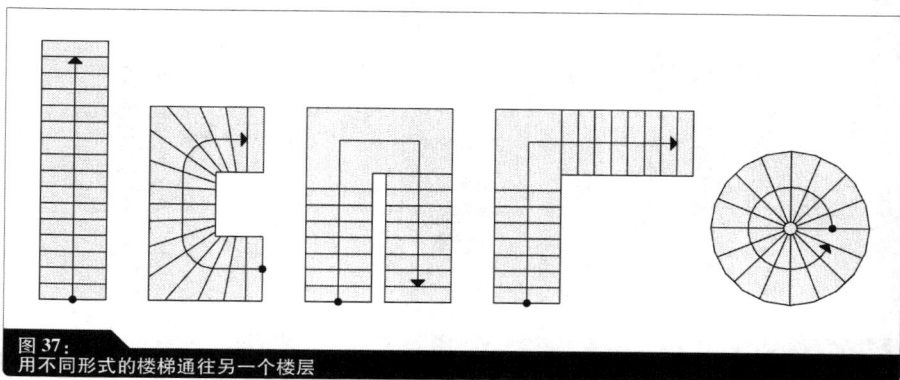

图37：
用不同形式的楼梯通往另一个楼层

建筑形式

　　前面关于住宅的章节主要是从空间和使用角度来讲解的。本章会从不同类型的角度、根据那些会对居住概念造成很大影响的特性对住宅进行分类。

独立式住宅

独立式和半独立式住宅

　　独立式住宅为单个家庭提供居所。根据当地的建筑规范，它们和邻居的房子之间需要保持一定的距离，而且还需要有很多的空间用作花园和基础设施。和城市建筑相比，这些面积会降低建筑密度（见图38）。

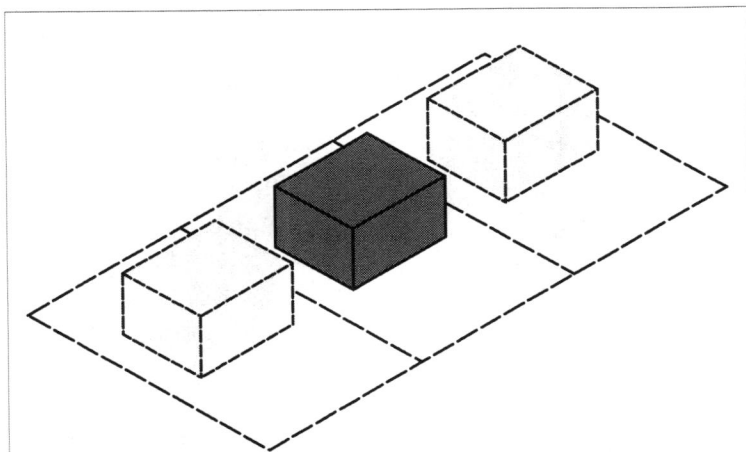

图38：
在城市开发项目中采用独立式和半独立式住宅要求有较大的土地面积，但是居住的品质能得到很大的提高

给住户的
暗示

　　从根本上来说，独立式住宅是单独的、而且最重要的是独立的居住变成可能。可以在设计中满足个人的风格和其他要求；有很大的创造空间。
　　这些单独的想法，以及包括以花园形式出现的私人户外空间，给设计提供了很多的发挥空间。

>◌

二层住宅　　　　　独栋住宅通常中会有两层：入口层通常包括日常使用的厨房、餐厅和起居室。比较私密的空间，比如说卫生间和卧室通常布置在二层。但是根据不同的要求和经济水平，增加标高可以让空间更加合理（见"基本内容，分区"和"空间创造"两节）。

平房　　　　　　　在平房中，不同功能的空间布置在同一层中。它们比较低矮的高度可以很好地跟乡村环境结合在一起。建筑在自己的空间内分区，可以通过凸出和凹进形成三维的感觉，并且形成与当地特征的联系。

　　　　　　　　　半独立式住宅在一栋房子中包括了两套住宅，每套属于一个家庭，它可以通过减少室外空间而降低造价（见图39）。首层通常是中线对称

半独立式

住宅

的，变化很小，但是也可以在同一个屋顶下建造两套不同的住宅。形状和材料要彼此统一，这样即使设计方法不同也能产生统一的效果。

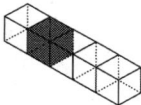

　　　　　　　　　把首层平面相同的住宅连成一排或者单个的居住单元组合起来可以形成排屋。它们通常被看作是比较省地的，因而也是比较经济的建设方式。它们一般占地很少，并且根据地块的宽度限制每户的开间（见图40）。

排屋　　　　　　　这些住宅有不同建造方式：住宅的前院可以是一条直线，一个圆形或者是多边形，可以是行列式，也可以是周边式布置。建设排屋可以实现比较高的城市密度，同时又能保持较高的居住品质，是非常值得推广的已经成形但密度又不是过高的城市开发模式。

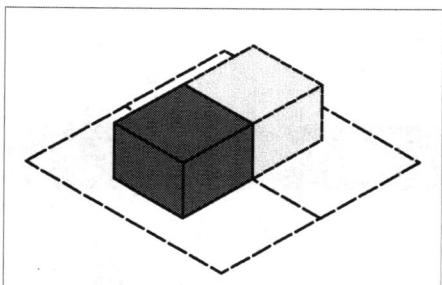

图39：
半独立式住宅在一栋建筑中包括了两套住宅

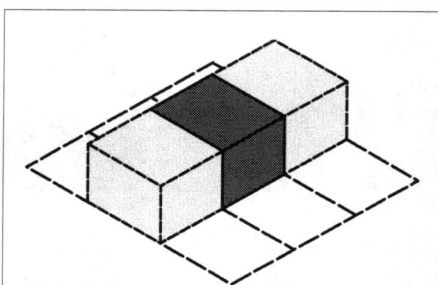

图40：
排屋通常占地很少。对于中间的单元来说，地块的宽度和住宅的宽度是相等的

跟独立式和半独立式住宅相比，采光条件受到了一定的限制：除了排屋首尾的两个单元之外，其他单元只有前后两个方向有自然采光。由于这个原因，有特定用途的不同房间的朝向必须进行很好的协调（见"基本内容，朝向"一节）。虽然排屋限制了个人的室外空间，但是可以采用新的方法。

**花园庭院
住宅**

花园庭院住宅是排屋的一种特殊形式。利用隔壁的山墙形成小型的私人庭院，而且这些庭院可以有很好的视线遮挡（见图41）。这个居中的户外空间为所有朝向它的房间提供了一个聚会的场所。因为当后面部分所有的窗户都能看到这个庭院的时候，也许会给这个私密空间的创造带来麻烦。

**"链式"住宅
P59**

为了打破立面的视觉形象，住宅通常会错落一下或者通过均匀的凸出和凹进形成链状布局（见图42）。

多层住宅

与独立式住宅不同，多层住宅是在一栋建筑中包括了若干个居住单元，彼此相邻或者在若干各楼层上彼此叠加。在面积较小的地块上开发多层住宅可以提高城市密度。

**街区周边
开发**

街区周边开发是一种封闭式的结构形式。一栋楼，或者是一系列单体建筑，围合了一个内庭院，从而创造出与外部不同的内部空间（见图43）。周边式开发可以做成不同的形状。矩形、圆形、弧线形、混合形，或者其他的几何形状都是可以的。根据尺寸和形状的不同，封闭的内庭院可以由其他建筑构成，也可以划分成另外的庭院。内庭院可以采取各种设计方法来加以处理，也可以有不同的用途。可以通过门、通道和入口等可能的进入方式形成公共的、半公共的和私密的效果。可以把它们设计成绿化空间、城市广场、花园和游戏场，也可以做成咖啡厅、商店和小型的公园。

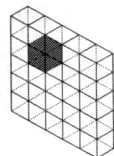

**对居住者的
暗示**

如果内部空间在很大程度上与外界隔离，那么它也阻挡了来自外部的干扰，比如说噪声、废气和视线，尤其是对于城市环境来说。这个内部空间通常具有半公共的特征，为居住者提供了绿化和游戏的休

提示：

要特别关注街区的转角，因为那里很可能成为自然采光和通风的死角。住宅中不同功能空间的朝向也应该根据太阳的位置进行认真分析，因为在不确定的情况下，必须考虑外部和内部空间孰轻孰重，以及具体的方位。

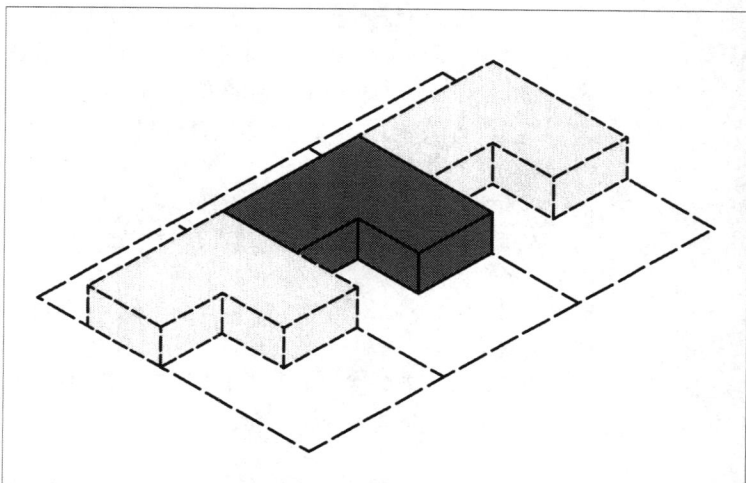

图 41：
花园提供了一个有很好视线保护的小型私人庭院

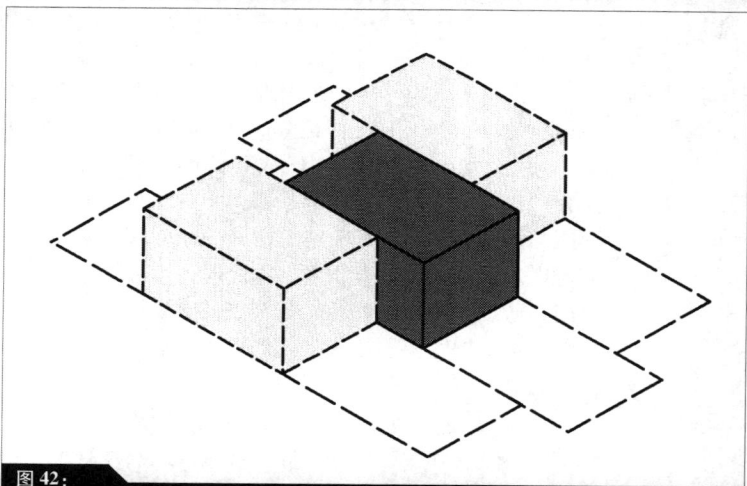

图 42：
因为它们的凸出和凹进，"链式"排屋为侧墙的自然采光增加了机会，而且可以在邻居之间形成遮挡视线的屏障

憩场所。起居室和卧室可以朝向这个安静的内部空间，如果可能的话，可以让入口和辅助用房朝向外部的城市空间，形成一个隔绝外部噪声和视线干扰的内部缓冲区。但是城市街区也可以简单地围合一个内部空间，通过首层架空、通道和其他的洞口，可以把这个内部空间

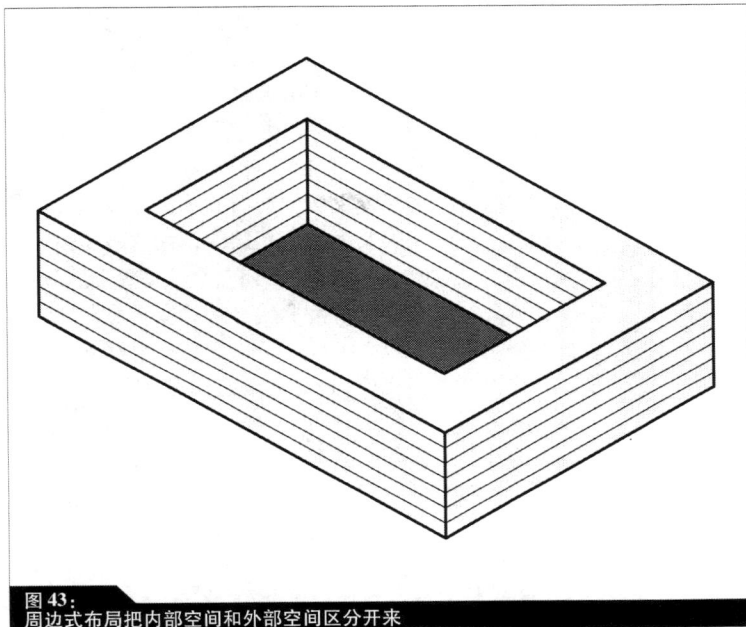

图43：
周边式布局把内部空间和外部空间区分开来

变成一个公共的场所，吸引城市生活进入其中。

联排住宅　　　这种结构形式指的是一系列排成行列式的单体建筑或者统一的街区。尽管它是一种非常简单的基本形式，但是彼此间不同的排列方式能够形成很多城市布局的变化。可以考虑采用平行的、垂直相交的或者斜线的布局。不同的长度和高度能够在一个复杂的结构中形成一种空间的联系。联排住宅不像周边布局那样可以清楚地把内部空间和外部空间隔开。如果采用平行的布局，楼间的空隙两端是开放的。这为自然采光和通风创造了良好的条件，但是可能会受到噪声、尘土和风的影响，尤其是在占地面积比较大的大型社区中。除此以外，楼间的空隙没有什么内在的品质，因为它们只有两面围合，而是通常能被整排住宅的视线所监视。还有一个办法是用其他与之成直角的楼把开放的端部给封闭起来。这样就可以有一个空间上比较封闭的内部环境，形成半公共的、安静的空间。

对居住者的　　　联排住宅可以布置成东西向或者南北向的模式，这样可以让不同
暗示　　　的功能很好地发挥作用（见"基本内容，朝向"一节）。与封闭的街区不同，联排住宅的角部不用进行特殊处理来解决通道、自然采光和通风

等问题。这意味着可以简单地采用标准的楼层平面。

街区和联排项目通常以连续的尺度为特征，在大面积的范围内留下它们城市开发理念的印记。

塔楼

与之相反，独立式的多层住宅通常会在单体建筑之间保持比较大的距离，它们通常会显得更高更长。这种比较大的尺寸意味着一组独栋住宅很难在人的尺度上形成连续的、空间上有变化的中间区域。

板楼

板楼是形式比较紧凑的线型结构，它的高度和长度通常都比较大（见图45左）。左如果与商业建筑相结合，即使是几栋板楼也很难形成内部区域，所以它们常常会形成不是那么容易设计的中间区域。

大型社区

板楼能够结合成大型社区，这样的社区可以是单独设计的，或者是以大规模开发为特征的（见图45中）。想要通过一个创造性的基本形式和组团来区分空间是不可能的，但是它可以在一定程度上加以运用，因为涉及的面积实在太大了。

点式建筑开发

点式建筑开发项目是由在一定范围内自由布置的塔楼所组成的（见图45右）。恰当的轮廓线设计能够在建筑上形成三维的形态，从而强调它们的垂直特征。不确定的组团形式意味着单体建筑和它们周围

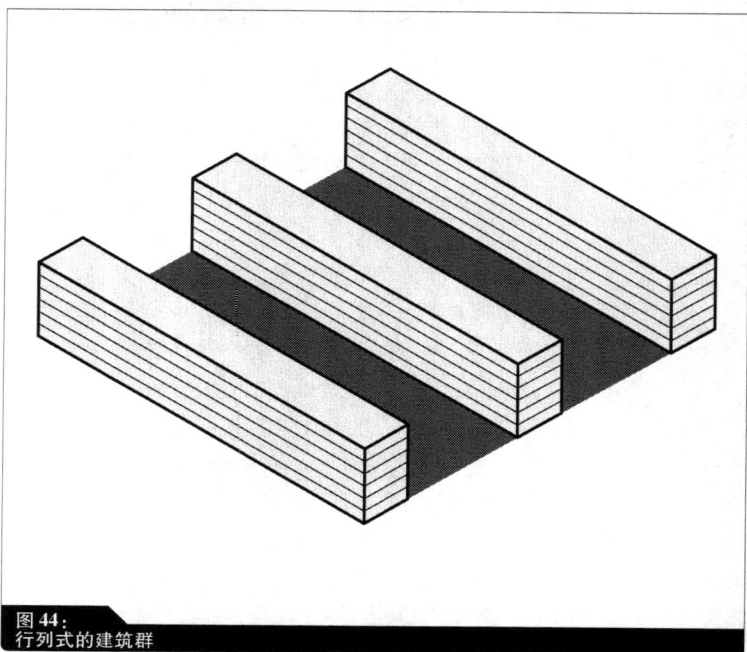

图44：
行列式的建筑群

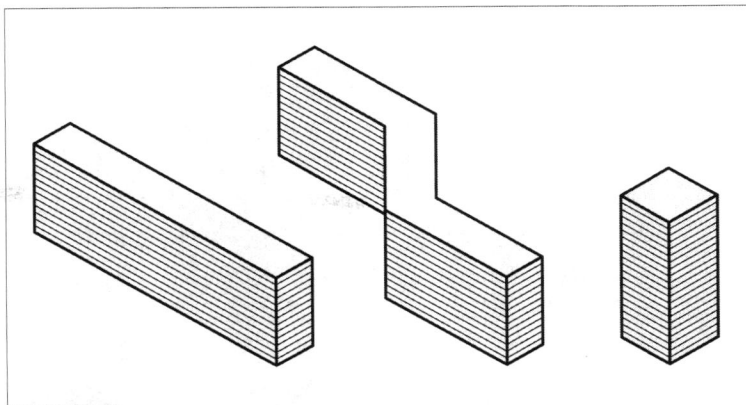

图45:
塔楼通常是高层的开发项目，在一栋建筑中容纳多个居住单元

的空间之间几乎没有什么关系，因为它们必须要离得很远，即使把它们布置在组团中也难以形成具有不同空间效果的中间区域。

对居住者的
暗示

塔楼把大量的居住单元放在一栋高层建筑内。通过首层平面和交通空间的优化设计能够形成紧凑而复杂的住宅建筑，能够在相对比较小的面积内解决很多人的居住问题。

起居空间的设计可以采取基本上比较个体化的方式，也可以采用标准化的形式。但是出于经济原因的考虑，大规模的住宅设计通常会采用统一的形式，具有可以在需要的时候建成不同类型的标准平面设计。公共的通道限制了个性化单元的出现，强调了建筑的共性。只有首层才能出现直接与每个家庭相连的花园。大大小小的阳台和屋顶平台也可以为多层建筑中的居住单元创造私人的室外空间。对于上面的楼层来说，朝向周围区域的开阔的视野也是一个优点。但是根据住宅所处位置的不同，它们会接触到特定的气候条件。高层住宅只能在有限的程度内享受到自然通风，因为它们位置较高，风力比较大。

举例：

勒·柯布西耶1947年设计的马赛公寓在一栋18层的建筑中布置了337个两层的公寓，每层楼的面积为138m×12m。建筑首层架空，所以它所覆盖的面积可以用作室外空间。第七层和第八层是各种商店和一个旅馆，屋顶设置了幼儿园、运动大厅和一个露天剧场。

标准化的系列产品把效率和经济性结合在一起，对很多人来说都很舒适。

"堆栈"式起居空间和其他的功能空间跟勒·柯布西耶的垂直城市模型是一致的。

通道

塔楼和行列式的独立或半独立式住宅都有它们自己的内部通道结构和单独的入口。这为建立私人的领域感提供了可能性，并且强调了这些住宅和结构形式各自的用途。从原则上来说，多层住宅也能实现这种类型的私人通道，但这通常意味着要有复杂的楼梯结构。下面会系统地介绍多层公寓的通道系统。它们包括诸如建筑入口之类的公共通道区域、居中的共用楼梯和电梯。

可以对不同的通道形式进行分类。我们可以区分出有直接通往楼梯的住宅和有走廊的住宅。对于多层住宅来说，根据建筑形式的不同，这两种类型的通道形式都是可行的。然而，走廊只有对那些建筑长度比较大、并且采用统一的通道系统的建筑来说才显得比较经济。

直接通往楼梯的模式

直接与楼梯相通的住宅楼会在中间的通道区附近布置一定数量的居住单元。每层的户数越多，整栋楼的经济性就越高，因为这样就有更多的居民能分享通道区，交通空间与使用空间的关系也比较好。

一梯一户式

一梯一户的住宅楼为每层的一户住宅提供通道，从根本上来说这是种不经济的形式，因为公用的楼梯只能为很少的住户服务，费用也就分摊到了少数几个居住单元身上（见图46）。通常来说这种形式的住宅会限制在4层以下，这样就可以不需要电梯。但是它在自然采光和通风以及首层平面的设计方面有很大的优势，因为每层其他的居住单元不会给它带来什么限制。

一梯两户式

一梯两户式住宅楼为每层的两户住宅提供通道，从而提高通道的利用率（见图47）。为了使结构更加合理，首层平面通常会采用轴线对称的格局，但是从房间的划分和大小来说，很难实现不同的户型。它的优点包括横向的穿堂风和至少两个方面的自然采光，这样可以让使

提示：

根据不同的建筑高度和相关规定，通道由踏步、楼梯的组合、电梯和紧急疏散楼梯决定。4层以上的建筑应设置电梯，否则很难保证居住单元的舒适性。地方建筑规范中关于高层建筑的条款规定了通道和防火要求。这些规定一定要在绘制最终的设计图纸之前就予以重视，而且它们通常会对通道和空间的结构产生深远的影响。

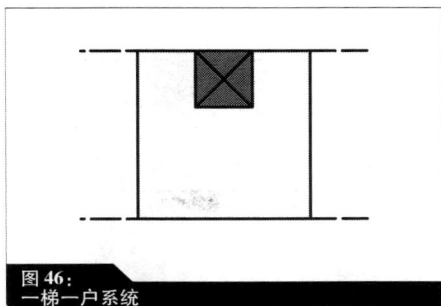

图46:
一梯一户系统

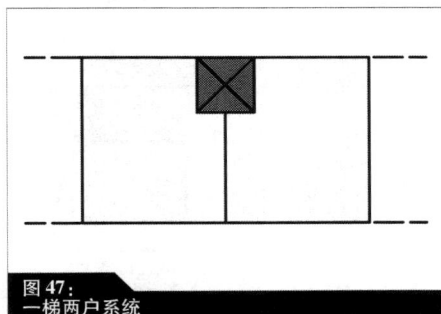

图47:
一梯两户系统

用空间有更好的朝向。

一梯三户式　　　　　一梯三户式住宅楼为每层的三户住宅提供通道（见图48）。根据不同使用者的需要可以设计不同大小和平面的户型，这样可以确保同一住宅楼中住户的多样化。根据单元划分的不同，它会在平面中形成单面朝向的户型，这会对不同使用空间的朝向带来局限性。横向的穿堂风也受到了限制。一梯三户式特别适合运用在建筑平面的角部。

一梯四户式　　　　　一梯四户式住宅楼可以同时为每层的四户住宅提供通道，所以是一种比较经济的形式（见图49）。单元的大小和户型可以有很多种变化。每层都可以有大户型和小户型。和一梯三户式住宅一样，它会造成单面朝向和很难形成穿堂风的问题。

提示：
　　一梯一户和一梯两户式很容易产生横向的穿堂风，通风效果良好。这意味着一户住宅可以拥有从建筑这一侧到另一侧的通风，在一个短的通风期内完成住宅空气的置换。短期的集中通风保证了空气的全部更换，并将热损失控制在一个最小范围。

115

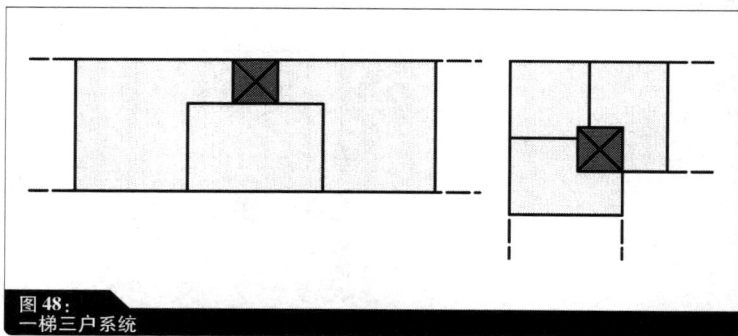

图 48：
一梯三户系统

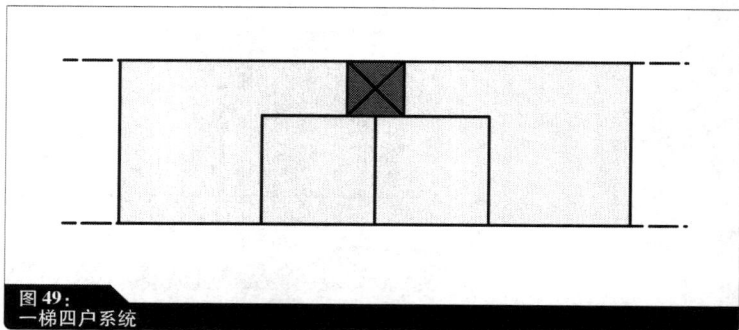

图 49：
一梯四户系统

点式住宅　　　　　　和塔楼一样，点式住宅不能像有直接通往楼梯的住宅类型那样系列化布置。它们把每层的居住单元组织在居中的垂直交通核周围。单元的数量和大小取决于建筑楼层面积的大小。从原则上来说，每个单元可以有两个朝向，这可以为合理的自然采光和通风创造条件——但是这仅限于每层不多于四个单元的平面。

中庭式住宅　　　　　　中庭式住宅是点式多层建筑的一种特殊形式（见图 50）。在这里，采光井能够为中间的住户带来自然光，它的高度是有限的，因为太阳光只有在一定条件下才能到达下面的楼层。中庭式建筑可以避免在建筑内部出现不受欢迎的黑空间，还可以在单元前面形成半私密的空间，给人比较宽敞的空间感觉。

走道式　　　　　　走道式住宅有服务于走道系统的垂直楼梯，这个走道系统分别为一定数量的居住单元提供水平通道。

　　　　根据建筑长度的不同，垂直通道可以居中布置或者局部布置。如果走道是在室内的，就称为内走道，如果在室外，就叫做外走道或者开放式走道（有顶盖的走道）。

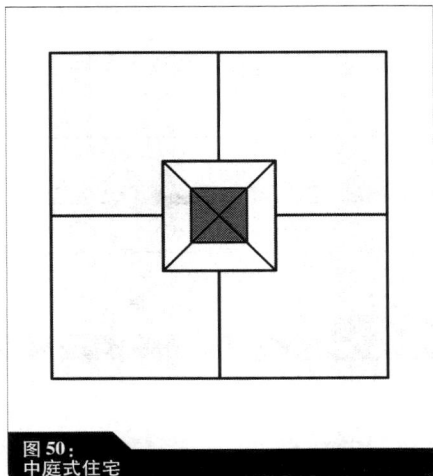

图50：
中庭式住宅

内走道　　　　　内走道的效率比较高，因为这样建筑的外皮可以完全用作居住单元的采光和通风（见图51）。反过来，内走道则几乎没有自然光。这就意味着这样的走道通常又黑又长，呆在里面会很不舒服，会产生不愉快的联想。要避免这种情况，可以在走道上比较合理的位置设计半公共的空间。这些空间与外立面相通，从而让自然光进入其中。这些空间中可以布置楼梯和电梯之类的垂直交通，或者设计成等候区或聚会点。

外走道/有顶盖的走道　　　　　由于它们是布置在建筑外墙上的，所以外走道没有采光的问题，而且也可以向室外敞开，也就是说，没有玻璃或者其他的隔断（见图52）。然而，对于高层建筑来说，还需要考虑气候条件。风和结冰限制了有顶盖的走道的功能。

竖向错落　　　　　从根本上来说，每层都可以有内走道或者外走道。但是这种布局方式意味着平面只能有一个朝向，因为走道遮挡了与建筑另一侧的直接联系，而且即使是采用外走道，也只能把次要的使用空间布置在走道一侧。在这种情况下，公寓和一居室的房子也许是个不错的解决办法。如果采用跃层式的格局，那么就只需要每两层或三层设置一个走道，无论是内走道还是外走道（见图53及"基本内容，空间创造"一节）。这样，一方面可以提高使用空间和交通空间之比，另一方面，有可能在建筑全进深方向形成穿堂风，而且至少有一层的使用空间可以有两个朝向。当然跃层户型不仅仅适用于走廊式住宅，直接从楼梯进入或者混合式的住宅都能从中受益。跃层户型是很好的把两个小户型放到一个大的单元中的办法，尤其是在老建筑中，这样就能满足现代的要求了。

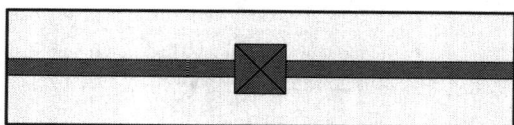

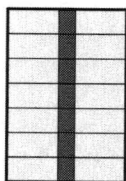

图 51：
内走道系统草图

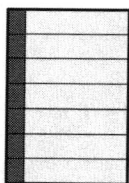

图 52：
外走道系统草图

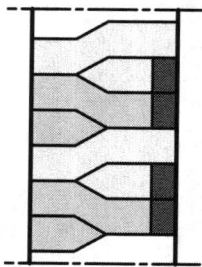

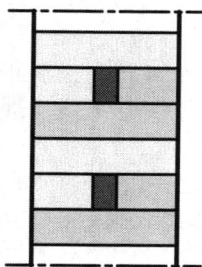

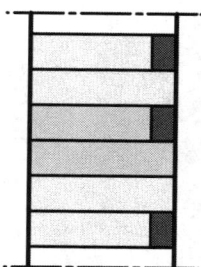

图 53：
竖向错落的走道能够提高跃层户型中起居空间和交通空间的比例。不同楼层的使用空间也可以划分成朝向两个方向的、没有的走道的空间

结语

> "建筑意味着设计生命的过程。"
>
> ——沃尔特·格罗皮乌斯

　　作为一条建筑原则，住宅建设有着无限的创造可能性。设计关键的生命过程跟现代的工艺和材料以及高度的舒适性和风格密切相关。起居空间所满足的基本需要最重要的特征就是促使人们把注意力集中到快速而高度发展的技术世界的本质上。在历史的进程中，住宅建设的目的和方法经历了巨大的变化，但是最重要的参数——人——几乎没有变过。我们仍然要设计出既适合当前又能面向未来的人类生活空间。新的方法和开放的思维跟回顾住宅的历史有着同样的意义。无数的理念、经验以及尝试可以跟新的想法、时尚的表现形式和新的工艺结合在一起。今天，住宅的品质与生态和能源密切相关。此外，社会人口的变化也会产生越来越大的影响力，要求有与之相适应的住宅出现。衡量所有解决这些变化的办法的标准是它们能否满足人类和他的基本需求，对于未来是否有意义。它不仅仅是一个满足必要的房子的问题，同时还是创造生活和居住的质量、不断地质疑我们熟悉的和标准化的住宅类型——还有最重要的，是创造而不是规定人的生活方式，是让他们能够创造性地开发和设计他们各自生活方式的环境的问题。

附 录
APPENDIX

参考文献
LITERATURE

Andrea Deplazes (ed.): *Constructing Architecture*, Birkhäuser Publishers, Basel 2005

Klaus-Peter Gast: *Living Plans*, Birkhäuser Publishers, Basel 2005

Manuel Gausa, Jaime Salazar: *Housing / Single Family Housing*, Birkhäuser Publishers, Basel 2002

Dirk Hebel, Jörg Stollmann (eds.): *Bathroom Unplugged*, Birkhäuser Publishers, Basel 2005

Ernst Neufert, Peter Neufert: *Architects' Data*, 3rd edition, Blackwell Science, UK USA Australia 2004

Périphériques / IN-EX projects: *Your House Now*, Birkhäuser Publishers, Basel 2003

Friederike Schneider (ed.): *Floor Plan Manual*, Birkhäuser Publishers, Basel 2004

Camillo Sitte: *The Birth of Modern City Planning: With a translation of the 1889 Austrian edition of his City Planning According to Artistic Principles*, Dover Publications, USA 2006

Klaus Spechtenhauser (ed.): *The Kitchen*, Birkhäuser Publishers, Basel 2005

Marcus Vitruvius Pollio: *Vitruvius: The Ten Books on Architecture*, Cambridge University Press, Cambridge UK 2001